Autonomous Weapons Systems and the Responsibility of States

Challenges and Possibilities

Lutiana Valadares Fernandes Barbosa

CRC Press is an imprint of the
Taylor & Francis Group, an **informa** business

Cover credit: Marcelo Fermandes

First edition published 2025
by CRC Press
2385 NW Executive Center Drive, Suite 320, Boca Raton FL 33431

and by CRC Press
4 Park Square, Milton Park, Abingdon, Oxon, OX14 4RN

CRC Press is an imprint of Taylor & Francis Group, LLC

Library of Congress Cataloging-in-Publication Data (applied for)

ISBN: 978-1-032-69232-6 (hbk)
ISBN: 978-1-032-69234-0 (pbk)
ISBN: 978-1-032-69236-4 (ebk)

DOI: 10.1201/9781032692364

Typeset in Times New Roman
by Prime Publishing Services

Foreword

By Prof. Aziz Tuffi Saliba

According to Greek mythology, Sisyphus was condemned to roll a massive boulder up a steep hill as a punishment from the gods for his deceitfulness. However, the rock would roll back down whenever he approached the top, forcing him to start over.

Like Sisyphus, we are always trying to catch up with the ever-evolving nature of technology. Just when we think we have grasped the implications of innovations, established proper regulations, and technology advances, we find ourselves back at the base of the hill.

As large language models awe us with their textual responses, self-driving cars navigate our streets, and autonomous weapons make decisions without significant human intervention, we are once again confronted with the potential and perils of Artificial Intelligence, sparking a worldwide demand for its regulation.

This book, evolved from meticulous research by Lutiana Valadares Fernandes Barbosa, delves deep into the transformative world of Autonomous Weapons Systems (AWS) and their implications within the framework of international state responsibility.

I first met Lutiana during her undergraduate years in law at the Federal University of Minas Gerais while teaching a Private International Law course. Her brilliance was evident, signaling a trajectory destined for remarkable achievements. Lutiana secured admission into some of Brazil's most competitive careers. Her unwavering commitment to justice and human rights led her to become a Federal Public Defender, a role she embraced with passion and dedication.

Her pursuit of knowledge continued as she furthered her studies with an LL.M. in Human Rights at Columbia University. Later, I had the privilege of guiding her as her academic supervisor during her PhD at the Federal University of Minas Gerais, which culminated in the research now presented in this book.

Lutiana provides an in-depth exploration of the complex interplay between AWS and the current international law on state responsibility, as encapsulated in the International Law Commission's Draft Articles on Responsibility of

States for Internationally Wrongful Acts. Readers are introduced to the inherent unpredictability of AWS, born from the intricacies of artificial intelligence. By examining AWS technologies' intrinsic unpredictability, the work reveals new challenges of attribution and accountability that require innovative approaches.

To address these gaps, Lutiana puts forth concrete and informed proposals. She calls for tailored attribution rules and a shift towards strict liability for states employing AWS in conflict. Her propositions pave an insightful path forward, from mandatory weapons reviews to minimum standards for human oversight. This cutting-edge prescription provides policymakers and nations with a crucial roadmap for promoting accountability amidst rapid technological change.

Facing technological advancements that challenge our traditional legal frameworks, clarity and direction become ever more crucial. Lutiana's comprehensive analysis acts as a compass, pinpointing the complexities and charting a course for a society navigating the waves of its technological creations. As we find ourselves perpetually at the foot of the Sisyphean hill of technological adaptation, this book serves as a guide and a testament to the human spirit's resilience and capability to innovate responsibly. It is a clarion call for collaborative efforts, urging legal scholars, scientists, engineers, policymakers, and nations to work together, placing ethics and accountability at the helm as we venture into new frontiers.

Foreword

By Prof. Daniele Amoroso

The contemporary landscape of armed conflict is undergoing a profound transformation, marked by the escalating integration of Artificial Intelligence (AI) and Autonomous Weapons Systems (AWS) into military arsenals. This transformative shift has ignited global discussions on the ethical and legal implications of autonomy in weapons systems, where international lawyers played a prominent role.

In the vast sea of literature resulting therefrom, Dr. Barbosa's book stands out for its focused approach. The book's exploration of State responsibility fills indeed a crucial gap in the existing literature. By undertaking a thorough examination of the legal implications and challenges associated with State responsibility in the realm of AWS, the book not only clarifies the current legal *status quo* but also engages with the topic also from a *de lege ferenda* perspective. Dr. Barbosa's keen navigation through the complexities of this relatively under-explored facet of the debate culminates in the proposition of pragmatic solutions, which are solidly grounded in established legal principles and ethical considerations, so offering a robust foundation for scholars and practitioners alike.

As we confront the profound implications of artificial intelligence on our societies, it is reassuring to know that scholars like Dr. Barbosa are researching these topics, by engaging in the collective academic enterprise to identify the path forward. Their commitment to understanding the ethical dimensions and legal intricacies of autonomous technologies contributes not only to scholarly knowledge but also to the broader societal imperative of ensuring responsible and accountable use of AI.

Acknowledgments

This book results from my 2019–2023 Ph.D. at the Federal University of Minas Gerais. I am extremely thankful to my advisor, Professor Aziz Tuffi Saliba, and my co-advisor, Professor Daniele Amoroso, for all the support throughout this journey. My gratitude to Professor Lucas Lima, to my colleagues from UFMG, especially Bárbara Solero, Bruno Biazatti, Mariana Ferola, and Sofia, to my colleagues from the Federal Public Defender's Office and the National School of the Federal Public Defender's Office, and my colleagues from the Ethics, Human Rights, and Artificial Intelligence study group (EDHIA), Ana Zago, Diego Silva, Fernanda Alves, Gustavo Macedo, and Viviane Dallasta, for all our debates.

I immensely thank the experts interviewed during my Ph.D., including Diego Mauri, Edson Prestes, Eugênio Garcia, Laura Bruun, Paul Scharre, and Rebecca Crootof. I only cite the interviewees who expressly allowed their nominal citation in the dissertation, but I am equally thankful to those with whom I conducted interviews and informal talks and who preferred to maintain confidentiality.

I have no words to thank my mother, Ana, and my father, Marcelo, for their lifelong support for education and for being my examples. Thanks to my husband Frederico, who made me see that Ph.D. research can be a leisure activity. Finally, my special gratitude to my son Daniel and my daughter Ana Luiza (born during the Ph.D.) for the inspiration, motivation, and wonderful moments of creative leisure.

Preface

This book seeks to answer whether the existing international legal regime on the international responsibility of States can effectively address breaches committed in the context of Autonomous Weapons Systems (AWS). To the extent that it cannot do so, it endeavors to map out the main gaps and some possible approaches to address them.

Part I sets the ground for the present discussion to guide this analysis. First, it provides a concept of AWS considering a degree of unpredictability as one of its inherent features. Next, it discusses the accountability gap around AWS and how the international community has put far more emphasis on individual responsibility rather than State responsibility.

Part II analyzes the challenges AWS pose to the regime governing State responsibility under international law, as codified in the Draft Articles on State Responsibility (ARSIWA). In this regard, it discusses attribution, breach of an international obligation, *tempus comissi delicti*, multiple States involved in a breach, force majeure, assurance of non-repetition, issues related to damage, the human-machine interaction and its impacts on State responsibility, responsibility for not using AWS, weapons review and the duty of due diligence.

Part III summarizes the challenges discussed in Part II in thirteen issues of concern grouped in three clusters and presents possible paths *de lege ferenda* to address them, achieved mainly by a paradigm shift in attribution and strict liability, among seven other more specific proposals.

The conclusion reached is that the current regime on the international responsibility of States is insufficient to deal with the new challenges AWS pose. *De lege ferenda*, the book argues for following the paths suggested in Part III. It also reflects on Parts II and III's findings and how many of the challenges AWS pose to State responsibility apply to other autonomous devices. Therefore, through the case study of AWS, this book also opens the broader discussion of the gaps in the international responsibility of States regarding autonomous device misdoings.

Lutiana Valadares Fernandes Barbosa

List of Abbreviations

AI	:	Artificial intelligence
AWS	:	Autonomous Weapons Systems
CCW	:	Convention on Certain Conventional Weapons
DoD	:	Department of Defense
ARSIWA	:	Draft Articles on Responsibility of States for Internationally Wrongful Acts
ECtHR	:	European Court of Human Rights
GGE	:	Group of Governmental Experts on Lethal Autonomous Weapons Systems
ICHR	:	Interamerican Court of Human Rights
ICRC	:	International Committee of the Red Cross
ILC	:	International Law Commission
HRW	:	Human Rights Watch
NGO	:	Non-governmental Organization
NATO	:	North Atlantic Treaty Organization
MHC	:	Meaningful Human Control
MoD	:	Ministry of Defense
NAM	:	Non-Aligned Movement
P-5	:	Permanent Members of the United Nations Security Council
UK	:	United Kingdom
US	:	United States
UN	:	United Nations
UNESCO	:	United Nations Educational, Scientific and Cultural Organization
UNGA	:	United Nations General Assembly
UN ICCPRC	:	United Nations International Covenant on Civil and Political Rights Human Rights Committee
UNILC	:	United Nations International Law Commission
UNSC	:	United Nations Security Council

Contents

Part II: International Responsibility of States and Autonomous Weapons Systems: Articles on the Responsibility of States For Internationally Wrongful Acts

Introduction

Once mere characters of science fiction movies, Autonomous Weapons Systems (AWS) are now part of the international reality. The Special Rapporteur on extrajudicial, summary, or arbitrary executions stated at the Human Rights Council "that more than 100 countries have military drones and more than a third are thought to possess the largest and deadliest autonomous weapons" (United Nations Office at Geneva, 2020). In 2021, the United Nations (UN) recognized for the first time (Conboy, 2021)[1] that an AWS, the STM Kargu-2 drone,[2] was used in the 2020 conflicts in Libya. The expert panel report to the Security Council affirms that "The lethal autonomous weapons systems were programmed to attack targets without requiring data connectivity between the operator and the munition: in effect, a true 'fire, forget and find' capability" (UNSC, 2021, p. 17).[3] Military experts affirm that AWS will likely be deployed in the ongoing Russian–Ukraine conflict if it does not end soon (Allen, 2022). Some even say Russia may have already used AWS (Kallenborn, 2022). In effect, we are at a moment in technological development where certain weapons might become autonomous long before previously anticipated (Garcia, personal communication, June 13, 2022).[4] Furthermore, it will be difficult to determine whether AWS was

---

1 For deeper comprehension, see appendix M.

2 For further information, see the STM website and its description of the STM KARGU2 https://www.stm.com.tr/en/kargu-autonomous-tactical-multi-rotor-attack-uav. According to the STM website "KARGU® is a portable, rotary wing attack drone designed to provide tactical ISR and precision strike capabilies for ground troops.
KARGU® is capable of performing fully autonomous navigation vis STM's unique flight control system. In addition, the platform is able to perform precision strike for low signature, beyond line of sight targets.
Precision strike mission is fully performed by the operator, in line with the Man-in-the-Loop principle.
The platform is capable of detecting and striking static or mobile targets with high precision during day and night conditions.
The KARGU system is comprised of the Attack Drone Platform and the Mobile Ground Control Station."

3 See Appendix D.

4 This is a citation from the author's Ph.D. dissertation. During an interview for the author's Ph.D. Eugênio Vargas Garcia, in his personal capacity, stated that:

(*Contd.*)

used, as changing from automated to autonomous mode might just be a matter of turning on the autonomy function 'button' (Scharre, personal communication, May 27, 2022).[5]

The concept of AWS is discussed in depth in Chapter 2, but a working definition is "A weapon system that, once activated, can select and engage targets without further intervention by a human operator" (US DoD, 2012). Most technological weapons today are semi-autonomous (US DoD, 2012), but a few cross the line into autonomy and can select and decide to engage targets without human intervention (Scharre, 2018, p. 46). Examples are the STM Kargu-2 mentioned above, the US Phanlax Close-in Weapon System, Russia's autonomous sentry robots (Del Monte, 2018, p. 163), and the Israeli Harpy,[6] sold to Chile, China, India, South Korea, and Turkey. Apparently, China has reversed-engineered its variant of the Harpy (Scharre, 2018, p. 47).[7] The leading futurist

---

"We are in a moment of technology that soon you will not be able to separate, and that certain weapons that are in the process of development can become fully autonomous, perhaps sooner than we thought." My translation of the original in Portuguese:

"estamos num momento da tecnologia que daqui a pouco você não vai ter muito como separar, e que certas armas que estão em processo de desenvolvimento podem se tornar totalmente autônomas, talvez antes do que pensávamos"

5 This is a citation from the author's Ph.D. During an interview for the author's Ph.D. Paul Scharre, in his personal capacity, stated that:

"One of the challenges even five or six years ago was that it wasn't really clear when technologically you'd be able to build an AI system that would be able to make targeting decisions correctly. I think we can say that actually we crossed that threshold. It doesn't mean that they've been used yet, but technologically, can we build an image classifier that can accurately identify different types of tanks, for example? Absolutely. Or other types of military equipment? Absolutely (…) I think we're starting in terms of the military operational component, to enter this sort of a gray area where we're starting to see military systems used in combat where the degree of autonomy is ambiguous."

6 For further information, see the IAI website and its description of the HARPY https://www.iai.co.il/p/harpy#:~:text=HARPY%20is%20an%20all%2Dweather,and%20search%20for%20radiating%20targets. It states that "HARPY is an all-weather day/night 'Fire and Forget' autonomous weapon, launched from a ground vehicle behind the battle zone. Programmed before launch to perform autonomous flight to a pre-defined 'Loitering Area,' in which they loiter and search for radiating targets. The HARPY loitering munition (LM) detects, attacks, and destroys enemy radar emitters, hitting them with high hit accuracy. HARPY effectively suppresses hostile SAM and radar sites for long durations, loitering above enemy territory for hours.

The launching process is controlled from the GCS and the HARPY LM is stored, transported, deployed, and launched from a canister mounted on a ground-based launcher normally mounted on trucks."

Some information on technical details and performance provided by IAI: "Technical Details: (a) Performance a.1) Communication Range: 200 Km a.2) Endurance: up to 9 Hours a.3) Speed: up to 225 Knots a.4) Low RCS: $<0.5m^2$ a.5) Maximum altitude: 15,000 Feet a.6 $<$ 1-meter precision strike with 16 kg warhead, (b) Performance b.1) Suppression/Destruction Enemy Air Defense Missions b.2) Autonomous operation b.3) State of the art Anti-Radiation seeker with wide RF coverage b.4) Vertical attack capability maximizing weapon lethality b.5) Abort attack in case of target shut down b.6) Longer range and extended loitering capability b.6) Continuous, persistent, lethal threat to enemy air defense systems.

7 For a further understanding on loitering munitions (see Atherton, 2021). For a deeper comprehension of AI and its military use in the United States, China, Russia, the United Kingdom, France, Israel, and South Korea (see Slijper, Beck and Kayse 2019).

on AI and AWS, Del Monte, believes that by 2050, States such as the United States (US), China, and Russia will use AWS with human-level autonomy (Del Monte, 2018, p. 113).

AWS are weapons that encompass strategic technology, represent significant military advantages, and are argued to be a necessary building block for military power. Thus, they are the focus of investment and development of many States, including the US, China (Allen, 2019),[8] and Russia (Konaev and Bendett, 2019). The President of Russia, Vladimir Putin, has stated that artificial intelligence (AI) is the future of humankind and that those at the forefront of AI will become the world's leaders. Among other states, the US, for instance, is investing billions of dollars in developing defense AI technologies (Slijper, Beck and Kayse, 2019).

Is humanity opening Pandora's box or using technology toward more humane and efficient armed conflicts? How do we deal with combat devices that do not feel compassion, fear, or remorse without losing our humanity? Do we human beings control our creations, or do (or will) they control us? (Myers, 2018).[9]

Robots acting autonomously using force and with decision-making power over life and death provoke questions in various knowledge areas. Legal rules are present in almost all branches of this discussion.[10] The central debates on AWS at the international level have been happening in Geneva under the auspices of the Group of Governmental Experts (GGE) on emerging technologies in the area of LAWS, created under the UN Convention on Certain Conventional Weapons (CCW). Despite diplomatic efforts, States have achieved little progress. The GGE operates through consensus, and leading military powers obstruct agreement on possible new regulation (Barbosa and Macedo, 2022).[11]

---

8 See also the opening paragraph of China's State Council's New Generation Artificial Intelligence Development Plan (AIDP) "AI has become a new focus of international competition. AI is a strategic technology that will lead in the future; the world's major developed countries are taking the development of AI as a major strategy to enhance national competitiveness and protect national security" (People's Republic of China, 2017).

9 In this sense, Del Monte states that "In the new reality, autonomous weapons will present a danger to humanity beyond their destructive potential. First, their autonomy may initiate a war that humans would potentially avoid via diplomacy. (…) Second, when autonomous weapon displace humans in harm's way, war may become more palatable, the probability of war increases. (…) Lastly, at the point of the singularity, superintelligence may view humanity as a threat, given our history of war and the release of malicious computer viruses. Armed with autonomous weapons, superintelligence may wage war on humanity, causing us to fall victim of own invention" (Del Monte, 2018, p. 113).

10 In this sense see for instance the Statement by the French General Secretariat for Defense and National Security "Robots and autonomous systems will be at the heart of the transformation of armed forces and the conduct of operations in the decades to come. This transformation will involve political, strategic, and technological aspects, and will need to be accompanied by normative and doctrinal support." My translation from the original in French "Les robots et systèmes autonomes seront bien au cœur de la transformation des armées et de la conduite des opérations dans les décennies à venir. Cette transformation comportera des volets politique, stratégique et technologique et devra être assortie d'un accompagnement normatif et doctrinal" (Secrétariat Général de la Défense et de la Sécurité Nationale, 2017, p. 196)

11 In this sense, Barbosa and Macedo stated that "The group's decisions are based on consensus, which often means facing insurmountable obstacles, given the tremendous interest from

(*Contd.*)

This book concentrates on one of the many challenges AWS pose to international law: the accountability hardships, focusing specifically on the international responsibility of States. Responsibility is vital, even in the ethically desired but politically unlikely event of a ban on AWS.[12]

The question we aim to answer throughout the book is:

Can the existing international legal regime on the international responsibility of States address breaches committed in the context of AWS? If not, what are the main gaps, and which possible approaches might bridge those gaps?

This book analyzes the strengths and difficulties of the existing legal regime of the international responsibility of States as foreseen in the Draft Articles on Responsibility of States for Internationally Wrongful Acts (ARSIWA) and suggests possibilities for suitable improvements and adaptations. To better understand the challenges of applying rules on State responsibility to AWS and trace venues to enhance such norms, we studied approaches by CCW's GGE, international organizations, documents by the five permanent members of the United Nations Security Council (P-5), Israel and South Korea, as well as doctrinal perspectives.

So far, human beings have been at the center of armed conflicts, and weapons have been passive objects (Pourcel, 2018, p. 17).[13] Nonetheless, AI paves the way for a fourth industrial revolution (Yang, 2018)[14] that might challenge the

leading military powers at the table. After adopting 11 general guiding principles in 2019, the governmental experts confronted a crisis of credibility for their failure to find paths to global governance of AWS and to make other substantial agreements. The situation was described as succumbing to the tyranny of the consensus rule since countries such as Russia have impeded progress" (Barbosa and Macedo, 2022).

12 Despite important advocacy led by NGOs for banning those weapons, which is morally very compelling, the political chances of success of a ban are very low. AWS imply military advantages, which lead inexorably to the rise of use of those weapons.
In this sense Saxon affirms that "The military advantages achieved by states and organized armed groups in possession of autonomous weapons technologies and capabilities will lead, inexorably, to the increasing use of autonomous functions and AWS." (Saxon, 2016, p. 208). Also, based on a 2017 survey by Ipsos, Del Monte states that "(...) banning autonomous weapons would be extremely difficult. However, given the results of IPSOS survey, there is an opportunity to regulate them" (Del Monte, 2018, p. 173).

13 The speech by NATO Secretary General Stoltenberg at the High-level NATO Conference on Arms Control and Disarmament also provides reflections on the new challenges posed by technological weapons: "Emerging technologies are also changing the game. Arms control has traditionally been about counting warheads, controlling numbers and distances. This is still relevant. But there are new threats on the field. Cyber, hypersonic glide, drones, autonomous weapon platforms, artificial intelligence, and biotech. They can all be weaponized. And in general, their military use is not constrained by international rules and regulations" (Stoltenberg, 2019).

14 We recall that the first industrial revolution was based on textile manufacturing and the innovation of the steam engine, the second industrial revolution on steel, electricity, and the automobile, the third triggered by the internet and mobile technologies.

(Contd.)

international balance (UK MoD, 2022, p. 10).[15] AWS has also been described as the third revolution in warfare after gunpowder and nuclear weapons (Garcia, 2016, p. 99). They represent a paradigm shift in accepting decision-making by devices (Pourcel, 2018, p. 77)[16] that challenge international accountability and call for developing new norms (Stoltenberg, 2019).[17]

AWS' actions are, at least to a certain degree, unpredictable due to interactions with the environment, their incapability to understand the broader context and the complexity of AI algorithms. Unpredictability, among other AI characteristics, such as a speed of decision-making that outpaces that of humans, raises challenges both to the international responsibility of States and individual criminal responsibility, as it may, for instance, lead to a break of the causal chain.

The gap in individual criminal liability is a relevant juridical problem outside the scope of this book since scholarship has focused on it (Dickinson, 2018, p. 19).[18] Still, few authors have discussed in-depth the international

---

See also (Abbott, 2020, p. 2) commenting that "Already impressive-sounding era titles such as the Fourth Industrial Revolution, the Second Machine Age, and the Automation Revolution are being used to describe the coming disruption."

15 See for instance the following excerpt from UK's Defence Artificial Intelligence Strategy "Adversaries and systemic competitors are investing heavily in AI technologies to challenge our defence and security edge. We have already seen claims that current conflicts have been used as test beds for AI-enabled autonomous systems, and we know that adversaries will use technology in ways that we would consider unethical and unsafe. Potential threats include enhanced Cyber and information warfare, AI-enabled surveillance, and population control, accelerated military operations and the use of autonomous physical systems. Non-state actors are seeking to weaponize advanced commercial products to spread terror and hold our forces at risk. As these case studies illustrate, this is not a hypothetical future but the here and now. As Hostile State and Non-State Actors build their AI capabilities, they will increasingly attempt to acquire key technologies and Intellectual Property from UK academia and Industry. This is a major threat. The UK must prepare to defend our most valuable technologies and capabilities, protect our universities from foreign interference, and safeguard industry, intervening where necessary" (UK MoD, 2022, p. 10).

16 In this sense Pourcel states that "In fact, AI endowment logically implies a paradigm shift and the consistent acceptance of ever greater autonomy in the way equipment operates and, consequently, the ability to make decisions within a standardized framework." My translation from the original in French "En effet, da dotation en IA implique par logique un changement de paradigm et lácceptation em cohérence dúne autonomie dáction de plus em plus forte des équipements dans leur mode de functionement et, par consequente, la capacite à prendere des decisions dans um cadre normé" (Pourcel, 2018, p. 77).

17 See for instance the excerpt of the speech by NATO Secretary General Jens Stoltenberg at the Institute for Regional Security and the Australian National University's Strategic and Defence Studies Centre: "We need to try to develop some norms for how we deal with these kind of new weapons systems, including autonomous weapons and other weapons, which will change the nature of conflict as fundamentally as the industrial revolution changed the nature of conflict before the First World War. So, we have just seen the beginning of a fundamental change of how weapon systems are working" (Stoltenberg, 2019).

18 Dickinson highlights that "A large portion of the scholarship on accountability for problems caused by autonomous weapons systems focuses on criminal responsibility and its limitations. In particular, a growing body of work highlights the difficulties in holding individuals criminally responsible for uses of unmanned and autonomous weapons that lead to significant violations of IHL" (Dickinson, 2018, p. 19).

responsibility of States regarding AWS (Amoroso, 2020, p. 131). We aim not to fill the criminal responsibility gap with State responsibility but to focus on the field of responsibility that has remained in the shade.[19] Therefore we will expose individual criminal responsibility only to the extent that it is necessary to elucidate the lack of focus on State international responsibility.

Overview of the Book

Besides the introduction, this book has three parts and a conclusion.

Part I sets the ground for the book's discussions. First, it provides a concept of AWS and addresses unpredictability as a feature of AWS. Next, it discusses the accountability gap AWS generate and how the international community has focused on individual responsibility rather than on State responsibility.

Part II analyzes the challenges AWS pose to State international responsibility, as stated in the ARSIWA, and presents thirteen issues of concern.

Part III groups these thirteen issues in three clusters and discusses possible paths *de lege ferenda.* Conclusions reflect on the findings of Parts II and III and how many of AWS's challenges to State responsibility apply to other autonomous devices.

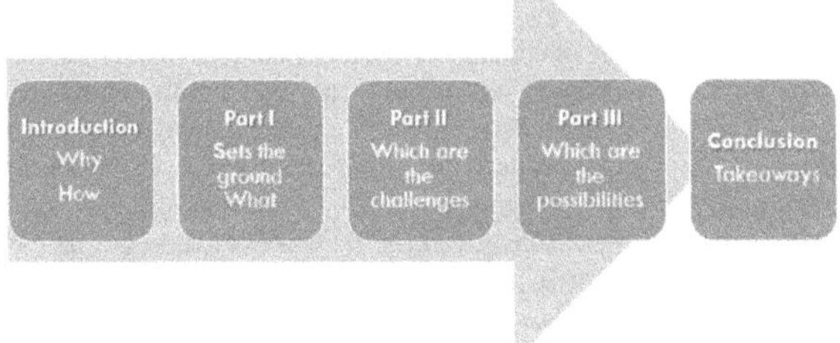

[19] Individual criminal responsibility does not exclude State international responsibility and can be enforced when appropriate, parallel to it. The problem arises when the individual criminal responsibility receives such a focus that State international responsibility is disregarded, or when there is an accountability gap in individual criminal responsibility and the responsibility of the State is ignored. In our view, studying State responsibility is necessary to enhance the international responsibility framework, but it does not solve or diminish the problem of the gap in individual responsibility. In this sense, we are aware of the relevant arguments presented by Daniele Amoroso stating that:

"The need to provide a proper legal response to serious violations of international law such as war crimes also explains, a fortiori, why proposals to fill the accountability gap by relying on forms of collective responsibility (such as State responsibility or corporate product liability) are largely unsatisfactory. In fact, the crucial, two-fold function of deterring the commission of international crimes and adequately retributing the offender for the harm done is peculiar to ICL and cannot be performed in the same way by collective responsibility, for the well-known reason that international crimes 'are committed by men, not by abstract entities, and only by punishing individuals who commit such crimes can the provisions of international law be enforced' " (Amoroso, 2017, pp. 21–22).

In an opposite view, Laura Dickinson states that: "Tort liability could potentially fill the void left by the failure of criminal doctrines to address many of the harms that could potentially result from the use of autonomous and semi-autonomous weapons" (Dickinson, 2018, pp. 23–26).

Methodological Remarks

This book is based mainly on bibliographic research. Interviews were also conducted.[20] They aimed to better understand experts' perspectives, verify other unforeseen gaps, and critically assess the results of the bibliographic analysis. A qualitative analysis was made of the interviews. The interviews were confidential, and citations were only made with the interviewee's authorization. Interviews cited in this book are a citation of the interviews cited in the author's Ph.D. dissertation.

The five permanent members of the United Nations Security Council (P-5) were chosen because they have the highest military spending worldwide (Statista, 2021), are at the forefront of the development of AWS, and have veto power in the UN Security Council, which has the central role of maintaining peace and security. Israel[21] and South Korea[22] were chosen because they are also among the States with the highest military spending worldwide (Statista, 2021) and are leading the development of AWS (Slijper, Beck, and Kayser, 2019, p. 5; Boulanin and Verbruggen, 2017, p. 95), especially regarding autonomous robotic sentry

[20] The researcher conducted (for the Ph.D. thesis that was adapted into this book) seven interviews, approved by Plataforma Brazil, using an unstructured questionnaire with leading experts in the fields of technology, international law, and security. For the questionary, see Appendix O.

[21] Slijper, Beck and Kayser state that "Israel is known as a tech-savvy nation, with a very innovative and burgeoning start-up scene," and "(...) the Israeli military deploys weapons with a considerable degree of autonomy. One of the most relevant examples is the IAI's Harpy loitering munition, also known as a kamikaze drone: an unmanned aerial vehicle that can fly around for a significant length of time to engage ground targets with an explosive warhead" (Slijper, Beck, and Kayser, 2019, p. 26).

[22] Slijper, Beck and Kayser observe that "South Korea did not make any statements at the 2018 Group of Governmental Experts (GGE) meetings in April and August, and it did not attend the CCW meeting of the High Contracting Parties in November 2018. However, South Korea did make statements in previous years. In April 2015, South Korea stated that "the discussions on LAWS should not be carried out in a way that can hamper research and development of robotic technology for civilian use," but that it is "wary of fully autonomous weapons systems that remove meaningful human control from the operation loop, due to the risk of malfunctioning, potential accountability gap and ethical concerns." In 2018, South Korea raised concerns about limiting civilian applications as well as the positive defence uses of autonomous weapons. "South Korea has repeatedly opposed any form of regulation of LAWS" (Slijper, Beck, and Kayser, 2019, p. 31). We add that South Korea in the CCW GGE 2022 was one of the authors of Principles and Good Practices on Emerging Technologies in the Area of Lethal Autonomous Weapons Systems Proposed by Australia, Canada, Japan, the Republic of Korea, the United Kingdom, and the United States CCW/GGE.1/2022/WP.2. The document states that "The High Contracting Parties to the Convention intend to: (a) Implement, as appropriate, these principles and good practices within each Party's respective national system; (b) Share, on a voluntary basis, their national policies and experiences relevant to the implementation of these principles and good practices, bearing in mind national security considerations and restrictions on commercial proprietary information; and (c) Keep these principles and good practices under review, and elaborate them by consensus as appropriate, while also continuing to consider and elaborate by consensus other possible measures and options related to the normative and operational framework on emerging technologies in the area of LAWS."

weapons (Boulanin and Verbruggen, 2017, p. 44).[23] Moreover, those States provide an overview of Western/non-Western perspectives and have China representing the global south States.

The analysis concentrated on CCW statements and documents of the P-5, Israel, and South Korea represents the present study's main limitation. Accounting for other States' perspectives is most relevant, and specific issues within this framework were thus considered. It is desirable to have a more equitable balance between the global south and global north States. Among the States studied, only China is a global south representative. However, since it is outside the scope of the present study to embrace all States, only those at the helm of AWS development were selected.

The limitations of analyzing mainly the P-5, Israel, and South Korea do not hamper the advantages of this choice. Those States have the biggest budgets on defense. They are at the forefront of AWS's development and are likely to provide a relevant overview of the developing debates on responsibility and AWS (Slijper, Beck, and Kayser, 2019, p. 5). In short, the chosen States' practices are deemed more significant than others since they are the most important States from the military point of view, as stated by the preamble of the CCW (CCW, 1983).[24] According to ICJ's decision in the North Sea Continental Shelf Case, some States' practices 'whose interests are especially affected' might be more relevant than others (ICJ, 1969, p. 43).[25]

The secrecy of defense information and difficulties in obtaining information regarding AWS, especially from governments, represents a challenge. Nonetheless, the selected States have presented their views at the CCW and published policies, reports, studies, or statements on AWS (US DoD, 2023; US DoD, 2012; UK MoD, 2022; UK, 2017; French Ministère des Armées, 2021).[26]

---

23 "Robotic sentry weapons are gun turrets that can automatically detect, track and (potentially) engage targets. They can be used as stationary weapons or be mounted on various types of vehicles. They resemble CIWSs but they use smaller caliber rounds and are usually employed as anti-personnel weapons" (Boulanin and Verbruggen, 2017, p. 44).

24 CCW's preamble states that: "Emphasizing the desirability that all States become parties to this Convention and its annexed Protocols, especially the militarily significant States" (CCW, 1983). The sentence demonstrates that militarily significant States develop an essential role in the context of the Convention.

25 See the ICJ's decision par. 74 "Although the passage of only a short period of time is not necessarily, or of itself, a bar to the formation of a new rule of customary international law on the basis of what was originally a purely conventional rule, an indispensable requirement would be that within the period in question, short though it might be, State practice, including that of States whose interests are especially affected, should have been both extensive and virtually uniform in the sense of the provision invoked; and should moreover have occurred in such a way as to show a general recognition that a rule of law or legal obligation is involved" (ICJ, 1969, p. 43).

26 Examples are: (1) United States, Department of Defense Directive n. 3000.09 (21 November 2012) and United States, Department of Defense Directive n. 3000.09 (25 January 2023), (2) United Kingdom, Joint Doctrine Publication 0-30.2, Unmanned Aircraft Systems (Published 12 September 2017 - Last updated 15 January 2018 - replaced Joint Doctrine 2/11), (3) Ministère des Armées, Defense Ethics Committee, Opinion on the Integration of Autonomy into Lethal

(*Contd.*)

Responsibility *de lege lata* and Liability *de lege ferenda*

There is some overlap between responsibility and liability; the difference is that responsibility focuses on the breach of an international obligation and does not necessarily require damage. Liability has compensation for damage as a linchpin and does not require a violation of an international obligation (Pemmaraju, 2000). The book has focused *de lege lata* on the analyses of the international responsibility of States, as stated in the ARSIWA. There is no general international liability regime, nor a specific liability regime related to AWS. Thus, we analyzed the existing framework of international responsibility to identify and address its gaps. *De lege ferenda*, measures proposed embrace both international responsibility and international liability.[27]

Weapons Systems (29 April 2021) (US DoD, 2023; US DoD, 2012; UK MoD, 2022; UK 2017; French Ministère des Armées, 2021).

27 In this sense, Crootof questions "(…) why shouldn't state responsibility extend to encompass all unanticipated harm to civilians resulting from actions attributable to a state?" (Crootof, 2016, p. 1402).

Part I
Setting the Ground for Discussing State Responsibility in the Context of Autonomous Weapons Systems

The following part provides a foundation for the subsequent discussions on AWS and the international responsibility of States. Firstly, the concept of AWS is introduced. Next, unpredictability as a feature of AWS is addressed. Finally, the accountability gap that AWS generate and how the international community has emphasized individual responsibility rather than State responsibility are discussed.

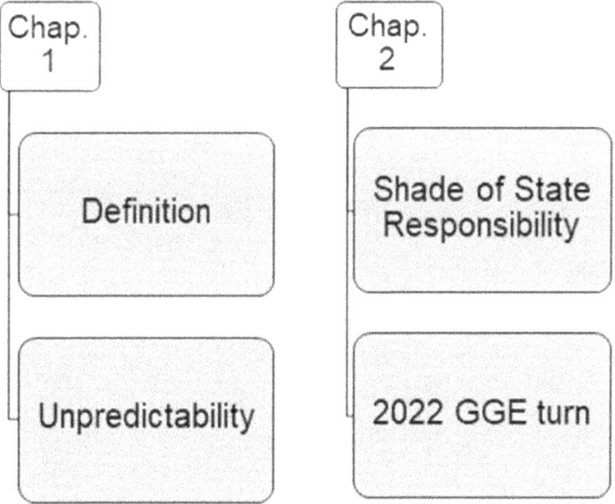

CHAPTER

1

Autonomous Weapons Systems: Concepts and Unpredictability

This Chapter begins by discussing whether a definition of AWS is necessary. It goes on to describe a scheme of the military decision-making process. States, international organizations, NGOs, and scholars' concepts of AWS are discussed, and a definition of AWS is provided. Finally, it considers the unpredictability of AWS both due to environmental interaction and the black box dilemma.

1.1 A Definition of Autonomous Weapons Systems (AWS)

In 2013, a report by the Special Rapporteur on extrajudicial, summary, or arbitrary executions, Christof Heyns, ignited the debate over the definition of AWS. His report[1] then recognized the divergence of concepts of AWS and proposed, as a starting point, that AWS are "robotic weapon systems that, once activated, can select, and engage targets without further intervention by a human operator. The important element is that the robot has an autonomous "choice" regarding selection of a target and the use of lethal force" (Heyns, 2013).[2]

Throughout the last decade, diplomacy has moved slowly. CCW GGE on LAWS meetings have occurred since 2014 informally and since 2017 formally[3], yet there is still no formal definition in the international arena (Davison, 2017; Taddeo and Blanchard, 2023). On the other hand, technology has moved much faster as AWS has been deployed on the battlefield (UNSC, 2021, p. 17).

---

1 For excerpts of the report, see Appendix B.

2 The report states that "37. While definitions of the key terms may differ, the following exposition provides a starting point. 38. According to a widely used definition (endorsed inter alia by both the United States Department of Defense and Human Rights Watch), the term LARs refers to robotic weapon systems that, once activated, can select, and engage targets without further intervention by a human operator. The important element is that the robot has an autonomous "choice" regarding the selection of a target and the use of lethal force" (Heyns, 2013).
The definition provided by Heyns was based on the 2012 US DoD and HRW positions, which will be discussed in the next sections.

3 For deeper comprehension, see Appendix A and Appendix C.

Experts, non-governmental organizations (NGOs),[4] international organizations, and States[5] diverge on the concepts (del Monte, 2018, p. 161; Scharre, 2018, p. 347). As the discussions evolved, definitions became politically loaded and could vary according to actors' interests, such as banning or permitting AWS (Scharre, personal communication, May 27, 2022).[6] While some focus on autonomy, others focus on human-machine interaction. Even if there is convergence over the broad framework, the meaning of the terms that compose the concept, differ—autonomy, for instance, is an ambiguous word (Ekelhof, 2017, p. 315). In this sense, Korea stated at the CCW GGE that it is unlikely that the group will agree on the definition of AWS soon (Korean Delegation, 2021). Furthermore, actors might change their concepts over time. Some even state that a definition is not necessary.

1.1.1 Is it Necessary to Define AWS?

It is worth questioning if we need an internationally agreed definition of AWS, as some of the legally binding disarmament instruments regulate different weapons use but do not define them, such as the CCW Protocol on Blinding Laser Weapons (CCW Protocol IV, 1998), the Treaty on the Non-Proliferation of Nuclear Weapons (NPT, 1970) and the Treaty on the Prohibition of Nuclear Weapons (Treaty on the Prohibition of Nuclear Weapons, 2017). Other treaties

4 IPRAW, for instance, stated at the CCW meeting in 2020 that "Although the United States defines what functions—that is, a weapon system that can select (search for, detect, identify, track or select) and attack (use force against, neutralize, damage or destroy) targets without human intervention." iPRAW recommends following this train of thought and fleshing it out further, for instance by excluding existing systems like stationary, anti-materiel weapons used solely to defend against incoming munitions (iPRAW, 2020).

5 For concepts of various countries see: (Soni and Dominic 2020, p. 12). See also, for example, the statement of the Brazilian delegation at CCW GGE 2020 "Thus, Brazil favors a workable, pragmatic definition of LAWS, that goes beyond the "technology-centric definitional approach". The concept proposed by ICRC-SIPRI1, elegant in its simplicity, is of great usefulness in this regard and should be adopted by the GGE: "Autonomous weapon system is any weapon system that once activated can select and attack targets without human intervention." For a more comprehensive definition of AWS, Brazil proposes the following addition: "An intelligent weapon system with autonomous operation mode (i.e., without human input after activation) capable of recognizing patterns in combat environments, and of learning to operate and make decisions regarding the critical functions of target identification, tracking, locking-on and engaging based on uploaded databases, acquired experiences, and its own calculations and conclusions" (Brazilian Delegation 2020a, p. 2).

6 This is a citation from the author's Ph.D. dissertation. During an interview for the author's Ph.D. Paul Scharre, in his personal capacity, stated that: "right now, definitions of AWS are politically very charged because you have a group of non-governmental organizations and some states calling for a preemptive legally binding treaty that would ban these AWS, but they haven't defined what an AWS is. And so instantly, that term becomes very loaded, and you see people on all sides of the issue engaging in some degree of strategic communication about how they define the term AWS because of this looming specter of the possibility that it could be banned. (…) When we worked on it in the DOD directive, none of this had happened yet. So it wasn't actually a factor at the time. (…) Our audience for the DOD directive was not diplomats at the United Nations because those discussions had not started yet. It was engineers inside the US defense department who would be building things, and we wanted it simple to communicate as clearly as possible to them."

provide definitions such as the Chemical Weapons Convention (Chemical Weapons Convention, 1997, art. 2), Protocol III to CCW on Prohibitions or Restrictions on the Use of Incendiary Weapons (Protocol III, on Incendiary Weapons, 1983, art. 1), and the Convention on the Prohibition of the Use, Stockpiling, Production, and Transfer of Anti-Personnel Mines and on their Destruction (Anti-Personnel Landmines Convention, 1997).

At CCW GGE, States such as China affirm that "Addressing the definition of LAWS is the key to negotiating any practical control measures" (People's Republic of China Delegation, 2022). On the other hand, Argentina, Costa Rica, Guatemala, Kazakhstan, Nigeria, Panama, the Philippines, Sierra Leone, Palestine, and Uruguay, in the "Roadmap Towards New Protocol on Autonomous Weapons Systems" hold a position with which we agree: "an exact technical definition of AWS is not required for the elaboration, development, and negotiation of any normative and operational framework" (Argentina and others, 2022a) on AWS considering that "autonomy exists on a spectrum and purely technical characteristics may alone not be sufficient to characterize AWS in view of rapid evolution in technology" (Argentina and others, 2022a). However, the document recognizes that a working definition is relevant as a starting point (Argentina and others, 2022a).[7] In short, our perspective is that a definition is fundamental, not for regulation, but to know what kind of weapons are addressed, even in broad terms (Crootof, personal communication, May 26, 2022).[8]

1.1.2 The Decision-Making Process and the Human Role: A Background to Definitions

To better discuss AWS's concepts, it is helpful to have two issues as a background, as many of the definitions are based on them: (a) the basic four-step OODA (Observe, Orient, Decide, Act) decision-making process often used in the military and weapons systems, and (b) the possible human roles in this decision-making process (Scharre, 2018, p. 43).

The OODA loop proposes that the decision-making process is essentially a four-step cycle:

---

7 The document further states that AWS "incorporates autonomy into the critical functions of selecting and engaging to apply force against targets, without human intervention. This means that a target is selected, and force is applied based on the processing of sensor data, rather than direct human inputs" (Argentina and others 2022a).
See also Draft Protocol VI (Argentina and others 2022b), which defines AWS as "weapon systems that incorporate autonomy into their critical functions of selecting, targeting, and engaging to apply force without human intervention."
Furthermore, the document calling for a Legally Binding Instrument on AWS (Argentina and others 2022c) which stated that such instrument shall include "Characterization of autonomous weapon systems."

8 This is a citation from the author's Ph.D. dissertation. During an interview for the author's Ph.D. Rebecca Crootof, in her personal capacity, stated that: "We need a definition of AWS to know what we're talking about, roughly, and to delineate different types, but that should be less for regulation and more to know what the conversation is about."

(1) Observe: Gather available information and identify a possible threat and its circumstances. In a weapons system, it means searching for a target(s);
(2) Orient: Analyze what was observed and design possible next steps. In a weapons system, it means detecting and selecting target(s);
(3) Decide: Choose among possible alternatives. In a weapons system, it means deciding to engage target(s);
(4) Act: Implement the decision taken in the previous step. In a weapons system, it means engaging a target.

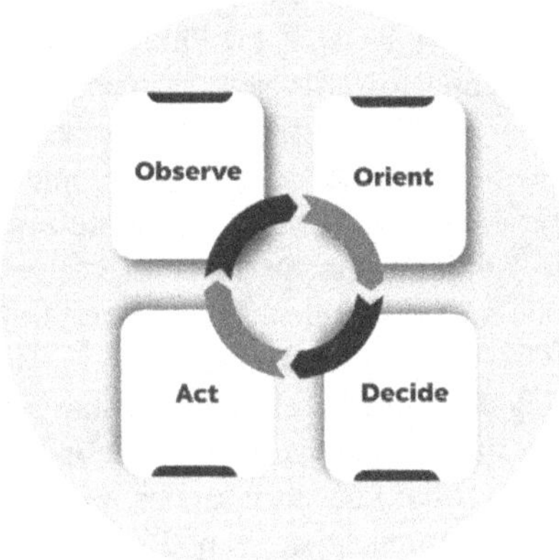

As regards human involvement in the autonomous/ automated weapons decision-making process, there are three main possibilities:

Human-in-the-Loop Weapons: The weapon develops a task and needs human action to continue (Scharre, 2018, p. 29), meaning it "can select targets and deliver force only with a human command" (Docherty, 2012).

Human-on-the-Loop Weapons: The weapon can observe, orient, decide, and act on its own. A human supervises and can intervene (Scharre, 2018, p. 29). A weapon "can select targets and deliver force under the oversight of a human operator who can override the robots' actions" (Docherty, 2012).

Human-out-of-the-Loop Weapons: The weapon can observe, orient, decide, and act on its own, without human interaction or input (Docherty, 2012). Humans cannot intervene in time to prevent the action (Scharre, 2018, p. 30).

1.1.3 AWS: A Spectrum of Definitions

The various AWS concepts[9]cover a broad technological spectrum.

9 Ekelhof, describes how controversial the definition process is at CCW "After two years of debating definitions of LAWS and concepts, such as autonomy, automation, self-governance, Artificial Intelligence and meaningful human control, the chairman of the 2016 informal meetings decided to plan a whole day of expert presentations and interventions focusing on 'Towards a Working Definition'. The response was, not surprisingly, controversial. Whereas the need for a shared definition or understanding of LAWS was shared by many, states did not agree on what a LAWS would constitute" (Ekelhof, 2017, 319).

At one extreme is the UK (UK, 2017), which requires AWS to have human-level intelligence, comprehend high-level intent, and make decisions without human oversight and control. Humans are out of the OODA loop. They claim that such systems do not exist yet and that they have no intention to develop them (UK, 2017).[10] The UK requires such a high threshold of AI that it has the effect of postponing and, perhaps, preventing the debate over AWS (Scharre, 2018, p. 110; Amoroso, 2020, pp. 15–17; UK Delegation, 2019).[11]

At the other end of the spectrum are countries like Russia, which even include remote-controlled devices (Russian Federation Delegation, 2018; Russian Federation Delegation, 2022),[12] meaning humans might be in, on, and out of the OODA loop.

France, China, the US, Israel, and South Korea are between the extremes.

---

10 United Kingdom, Joint Doctrine Publication 0-30.2, Unmanned Aircraft Systems – Published 12 September 2017 – Last updated 15 January 2018 substituted Joint Doctrine 2/11. "The UK does not possess fully autonomous weapon systems and has no intention of developing them. Such systems are not yet in existence and are not likely to be for many years, if at all."

11 Paul Scharre highlights that "This definition shifts the lexicon on autonomous weapons dramatically. When the UK government uses the term "autonomous system," they are describing systems with human level intelligence that are more analogous to the "general AI" described by US Deputy Defense Secretary Work. The effect of this definition is to shift the debate on autonomous weapons to far-off future systems and away from potential near-term weapon system that may search for, select, and engage targets on their own- what others may call "autonomous weapons" (Scharre, 2018, p. 110).
Amoroso, explains the disagreement within the UK government: "Even the UK's House of Lord's Select Committee on AI, has stated that such a concept "is clearly out of step with the definitions used by most other governments" and needs to be reframed" (Amoroso, 2020, pp. 15–17). Amoroso cites the UK parliament website: "**AI in the UK: ready, willing and able? Summary of conclusions and recommendations**".
At CCWGGE, the UK advocated for a characterization of AWS that focuses on methods to maintain human control.
"In summary, the UK believes that a technology-agnostic approach to characterizing LAWS which focuses on methods for maintaining human control is likely to be the most productive and robust in the face of future technological developments. We recognize that there are multiple decisions made by humans throughout a weapon's lifecycle which combine to enable human control to be achieved and tailored to the demands of the specific task and operational environment" (UK Delegation, 2019).

12 At the CCWGGE, Russia stated that "According to Russian experts, the working definition should meet the following requirements: (a) it should contain the description of the types of weapons that fall under the category of LAWS, conditions for their production and testing as well as of their use; (b) the wording should not be limited to the existing understanding of LAWS, but also take into consideration the possibility of their future development; (c) it should be universal in terms of the understanding by the expert community, including scientists, engineers, technicians, military personnel, lawyers and ethicists. While elaborating the definition of LAWS, one should avoid making hasty decisions "cementing" technological advancement" (Russian Federation Delegation, 2018).
For instance, the Russian Ministry of Defence uses the following working definitions: (a) "Autonomous weapons system – an unmanned piece of technical equipment that is not a munition and is designed to perform military and support tasks under remote control by an operator, autonomously or using the combination of these methods" (Russian Federation Delegation, 2018).
In the same sense (Russian Federation Delegation, 2022).

France also requires technology far from the current state of the art, defining AWS as fully autonomous lethal weapons systems, which means no human involvement. Humans are out of the OODA loop. France's definition precludes such weapons from existing so far (French delegation, 2018; Wareham, 2020, French Ministère des Armées, 2021, p. 4),[13] but is not as extreme as the UK.

China also leans towards a high technological position as it places humans out of the OODA loop and requires machine learning and the impossibility of a human aborting the action (People's Republic of China, 2018).[14]

The US, like China, requires the device to be capable of pursuing the task without human intervention but allows human override of the device. Humans are, therefore, on or out of the OODA loop. The US Department of Defense (DoD) defines AWS as:

> *A weapon system that, once activated, can select and engage targets without further intervention by an operator. This includes, but is not limited to, operator-supervised autonomous weapon systems that are designed to allow operators to override operation of the weapon system, but can select and engage targets without further operator input after activation (US DoD, 2023 p. 21).*[15]

---

13 The French Delegation stated at CCW GGE that "It must first be recalled that although autonomous technologies are growing quickly, LAWS are still very much a long-term initiative: there are currently no fully autonomous lethal weapons systems" (French Delegation, 2018).
One year before, it also stated that "Formulating a working definition would be useful to frame the discussions of the GGE and agree on possible measures to address the issue of LAWS. Recalling that LAWS do not exist yet and given the wide range of possible definitional approaches we propose agreeing on a preliminary working definition whereby LAWS are defined as fully autonomous lethal weapon systems. Systems such as remotely piloted and automated systems (e.g. conventional charges exploding with a set timer), tele-operated (e.g. drones), automated missile defence systems, torpedoes, guidance and navigation systems, surveillance and detection systems are not considered as LAWS" (French Delegation, 2017).
Human Rights Watch stated that "It considers killer robots to be "weapon systems that have no human supervision once they are activated" (Wareham, 2020).
The French Ministère des Armées distinguishes lethal autonomous weapon systems (LAWS), "capable of changing their rules of operation and therefore are likely to depart from the employment framework initially defined" from partially autonomous lethal weapon systems (PLAWS), which are under human control and "whose decision-making functions are defined according to the robotics meaning of decision-making autonomy, i.e. within a specific framework of action" (French Ministère des Armées, 2021, p. 4).

14 At CCW, China stated that: "In our view, LAWS should include but not be limited to the following 5 basic characteristics. The first is lethality, which means sufficient payload (charge) and for means to be lethal. The second is autonomy, which means absence of human intervention and control during the entire process of executing a task. Thirdly, impossibility for termination, meaning that once started there is no way to terminate the device. Fourthly, indiscriminate effect, meaning that the device will execute the task of killing and maiming regardless of conditions, scenarios and targets. Fifthly evolution, meaning that through interaction with the environment the device can learn autonomously, expand its functions and capabilities in a way exceeding human expectations" (People's Republic of China Delegation, 2018, p. 1).

15 See also the former definition US DoD, 2012, p. 13 "A weapon system that, once activated, can select and engage targets without further intervention by a human operator. This includes human-

(Contd.)

South Korea distinguishes AWS from remotely controlled weapons (Republic of Korea Permanent Mission, 2014), and Israel emphasizes that humans, not machines, make decisions (Israel Permanent Mission, 2018).[16]

Both the UK (2011) and the US DoD (2012) directives were issued prior to the United Nations' debates on AWS and whose target audience was engineers inside the DoD and not diplomats at the United Nations (Paul Scharre, personal communication, May 27, 2022).[17] Other definitions, such as ICRC and Stop Killer Robots, were formulated after the UN debate began and are politically loaded with the interests the organizations or States convey, such as banning AWS or allowing their use.

The autonomy feature gravitates between two ends of a spectrum: autonomy as a human feature linked to free will and autonomy as task autonomy. The illuminist model that understands autonomy as human-like free will is the one that the UK's definition leans toward, requiring a weapon with human-level intelligence. The model of task autonomy, meaning the capability to develop tasks, is the one China's, Russia's, and the US's definitions gravitate towards. The difference between task autonomy, applicable to AI, and personal autonomy, which is the illuminist framework of autonomy grounded on free will, is further discussed in Chapter 4.

The International Committee of the Red Cross (ICRC) and the Campaign to Stop Killer Robots also embrace this task-oriented view of autonomy and adopt definitions along the lines of the US DoD concept. The ICRC adopts a broad

---

supervised autonomous weapon systems that are designed to allow human operators to override operation of the weapon system, but can select and engage targets without further human input after activation."

During an interview for the author's Ph.D., Diego Mauri, in his personal capacity, stated that: "The DOD definition is quite synthetic but captures the essential feature of this technology. On one hand the idea of selecting engaging targets. This is a technology that can take lethal action against targets in general. (…)the critical function idea, coupled with the autonomy ideas, the idea that they, that these weapons system can do without human intervention" (Mauri, personal communication, July 5, 2022). This is a citation from the author's Ph.D. dissertation.

16 Mrs. Maya Yaron, Minister, Counsellor Deputy Permanent Representative to the CCW on LAWS, stated that "As far as terminology is concerned, that means that LAWS should not be regarded as "deciding" anything. Humans are always those who decide, and LAWS are decided upon" (Israel Permanent Mission, 2018).

17 This is a citation from the author's Ph.D. dissertation. During an interview for the author's Ph.D., Paul Scharre, in his personal capacity, stated that:
"right now, definitions of AWS are politically very charged because you have a group of non-governmental organizations and some states calling for a preemptive legally binding treaty that would ban these AWS, but they haven't defined what an AWS is. And so instantly, that term becomes very loaded, and you see people on all sides of the issue engaging in some degree of strategic communication about how they define the term AWS because of this looming specter of the possibility that it could be banned. (…) When we worked on it in the DOD directive, none of this had happened yet. So it wasn't actually a factor at the time. (…) Our audience for the DOD directive was not diplomats at the United Nations because those discussions had not started yet. It was engineers inside the US defense department who would be building things, and we wanted it simple to communicate as clearly as possible to them" (Paul Scharre, personal communication, May 27, 2022).

perspective and considers AWS as a weapon "that has autonomy in its '**critical functions**,' meaning a weapon that can select (i.e. search for or detect, identify, track) and attack (i.e. intercept, use force against, neutralize, damage or destroy) targets without human intervention" (ICRC, 2015).[18] The Campaign to Stop Killer Robots, a coalition of NGOs focused on banning AWS, defines it as "(…) weapons systems that would select and engage targets based on the sensor inputs, that is, systems where the object to be attacked is determined by sensor processing, not by humans." (Docherty 2019b, 2) This means that critical stakeholders "converge around this formulation of central properties of "autonomy": the technologically most advanced military power, the main international humanitarian organization, and the coalition of NGOs advocating for a ban on AWS" (Amoroso, 2020, p. 17). Leading scholars also adhere to those definitions, at least in broad terms.[19]

18 ICRC expresses the reason for adopting a broad definition: "The advantage of this broad definition, which encompasses some existing weapon systems, is that it enables real-world consideration of weapons technology to assess what may make certain existing weapon systems acceptable – legally and ethically – and which emerging technology developments may raise concerns under international humanitarian law (IHL) and under the principles of humanity and the dictates of the public conscience" (ICRC, 2016).

In 2020, the ICRC stated at the CCW GGE meeting that "Autonomous weapon systems, as the ICRC understands them, select and apply force to targets without human intervention. To varying degrees, the user of the weapon will know neither the specific target nor the exact timing and location of the attack that will result. This raises serious concerns from a humanitarian, legal and ethical perspective, in particular the risk of losing human control over the use of force" (ICRC, 2020b).

Dickinson comments that "By contrast, the International Committee for the Red Cross (ICRC) does not clearly distinguish among multiple categories of autonomy, instead describing an autonomous weapon as one "that is able to function in a self-contained and independent manner although its employment may initially be deployed or directed by a human operator" and that can "independently verify or detect a particular type of target object and then fire or detonate"" (Dickinson, 2018, p. 39).

19 See for example, Paul Scharre defines AWS based on their ability to complete the decision-making process, "searching for, deciding to engage, and engaging targets on their own." He highlights that both supervised and fully AWS are built and deployed by humans, who are "involved in the broader process of designing, building, testing and developing weapons" (Scharre, 2018, p. 52).

Crootof also provides a definition that highlights AWS's capability of pursuing a decision-making process. AWS "is a weapon system that, based on conclusions derived from gathered information and preprogrammed constraints, is capable of independently selecting and engaging targets" (Crootof, 2015, p. 1854).

Gracia states that "Autonomous weapons select and engage targets without human intervention. They might include, for example, armed quadcopters that can search for and eliminate people meeting certain predefined criteria, but do not include cruise missiles or remotely piloted drones for which humans make all targeting decisions" (Garcia, 2016, p. 99).

Ford affirms that "A common definition of an autonomous weapon used in the legal literature is a weapon that can select and engage a target without human involvement. While this definition is commendably succinct, it raises significant questions: What does it mean to "select"? How and when can systems operate "without human involvement" if all systems are programmed by humans?" (Ford, 2017, p. 417).

Dickinson, 2018, p. 8 "They maintain that an essential aspect of autonomy is the ability of weapons systems to engage in "selection among" targets" (Dickinson, 2018, p. 8).

Slijper, Beck and Kayser, affirm that "Lethal autonomous weapon systems are weapons that can

(*Contd.*)

Task autonomy grasps what AWS are capable of performing tasks without further human intervention. It is also present in the North Atlantic Treaty Organizations' (NATO, 2020)[20] and the Institute of Electrical and Electronics Engineers' (IEEE, 2021a)[21] definitions of autonomous systems (both have not yet defined AWS). NATO and the IEEE also recognize adaptability based on evolving situational awareness and the likelihood of an unpredictable decision-making process. The Future for Life Institute also envisions autonomy as a task-oriented and explicit AI feature by defining AWS as "weapon systems that use artificial intelligence (AI) to identify, select, and kill targets without human intervention" (Future for Life Institute, n. d.). The AI feature will be further analyzed in the next section.

1.1.4 A Working Definition of AWS for the Present Research

In this section, we will provide a working definition of AWS that, in the author's view, better presents its features.[22] The aim is not that this definition prevails but to provide a broad framework of the weapons included when discussing State responsibility for AWS breaches.

Most of the discussed definitions converge that AWS can select and engage targets without human intervention, can gather and analyze information, and use force autonomously.[23]

---

select and attack individual targets without meaningful human control. This means the decision on whether a weapon should deploy lethal force is delegated to a machine. This development would have an enormous effect on the way war is conducted and has been called the third revolution in warfare, after gunpowder and the atomic bomb" (Slijper, Beck and Kayser, 2019).

20 NATO's terminology database defines an autonomous system as "Pertaining to a system that decides and acts to accomplish desired goals, within defined parameters, based on acquired knowledge and an evolving situational awareness, following an optimal but potentially unpredictable course of action" (NATO, 2020).

21 The IEEE's Ontological Standard for Ethically Driven Robotics and Automation Systems defines autonomous systems as "(…) capable of performing tasks and behaviors with a high degree of autonomy making informed decisions without external direction and with the ability to adapt to changing conditions, knowledge, and constraints" (IEEE, 2021a, p. 12).

22 For other proposals of definitions, see for instance (Crootof, 2015) and (Taddeo and Blanchard, 2023).

23 On the scholarship, see for instance: (Garcia, 2016, p. 99) "Autonomous weapons select and engage targets without human intervention. They might include, for example, armed quadcopters that can search for and eliminate people meeting certain predefined criteria, but do not include cruise missiles or remotely piloted drones for which humans make all targeting decisions"; (Ford, 2017, p. 417) "A common definition of an autonomous weapon used in the legal literature is a weapon that can select and engage a target without human involvement. While this definition is commendably succinct, it raises significant questions: What does it mean to 'select'? How and when can systems operate 'without human involvement' if all systems are programmed by humans?"; (Dickinson, 2019, p. 8) "Commentators Kenneth Anderson and Matthew Waxman have observed that, despite significant differences, these definitions share a common view of what makes a weapon system 'autonomous': it is a matter of whether a human operator realistically is able to override an activated machine in the core function of target selection and engagement." They maintain that an essential aspect of autonomy is the ability of weapons systems to engage in 'selection among' targets" (Slijper, Beck and Kayser, 2019). "Lethal autonomous weapon systems are weapons that can select and attack individual targets without meaningful human

(*Contd.*)

It is AI that makes AWS's decision-making process possible because it enables the devices to pursue the entire OODA loop autonomously, especially when analyzing data and opting for a course of action. AI "involves computational technologies that are inspired by – but typically operate differently from – the way people and other biological organisms sense, learn, reason, and take action" (IEEE, 2019). AI is a problem-solving computer program capable of decision-making. It is grounded on inferential reasoning and can solve problems and decide on situations where information is incomplete or uncertain. AI embraces a broad spectrum: from previously programmed rules, which the program chooses based on inference or evaluation of options, to modern programs that operate through machine learning and create their scoring formula. In the latter case, programmed rules are not on how to solve a problem but on learning from data (Bathaee, 2018, p. 898). The concepts of AWS presented in the previous sections vary in this spectrum of AI. While simple algorithms suffice for some, others require human-like intelligence.

This book disagrees with concepts such as the UK's that require such a high technological threshold that has the side effect of postponing or perhaps preventing relevant juridical debates. Concepts such as the US DoD, Stop Killer Robots, ICRC, and Future for Life Institute are far more useful in providing impetus for AWS discussions.

Nonetheless, the definitions mentioned above are not explicit regarding a crucial element: AWS can make, within certain limits, discretionary decisions, which means they decide when, how, and which targets to attack (Petman, 2017, p. 70).[24] The defining feature of AWS is the freedom to make choices once initiated (Scharre, 2018, p. 50).[25] This book adopts a modified version of the

---

control. This means the decision on whether a weapon should deploy lethal force is delegated to a machine. This development would have an enormous effect on the way war is conducted and has been called the third revolution in warfare, after gunpowder and the atomic bomb. The function of autonomously selecting and attacking targets could be applied to various platforms, for instance a battle tank, a fighter jet or a ship."

24 The important feature of discretion is raised by Petman, "There are at least two characteristics that define the notion of autonomy in the specific context of AWS. First is the ability to operate independently and engage targets without being programmed to specifically target an individual object or person. This includes the capability to react to a changing set of circumstances, and requires that the rules of IHL be "translated" into machine code. The second, interrelated, aspect is the capability to make discretionary decisions. The actions of AWS are therefore, in contradistinction to automated systems, predictable only within the range that they were programmed. The definitions provided by both the DoD and HRW do not provide this crucial element. Deciding which targets to engage, as well as how and when to carry out an attack, would be left to the AWS's software that is programmed to deal with a myriad of situations and changing set of circumstances" (Petman, 2018, p. 70).

25 Scharre describes freedom as a defining feature of AWS: "It is freedom, not intelligence, that defines an autonomous weapon. Greater intelligence can be added into weapons without changing their autonomy. To date the target algorithms used in autonomous and semiautonomous weapons have been fairly simple. This has limited the usefulness of fully autonomous weapons, as militaries may not trust giving a weapon very much freedom if it isn't very intelligent. As machine intelligence advances, however, autonomous targeting will become technically possible in a wider range of situations" (Scharre, 2018, p. 50).

DoD's concept within a framework that stresses its discretion and AI. For the purposes of the present research, AWS are:

Weapons systems that operate through AI and can discretionarily select and engage targets without further intervention by a human being.

According to the provided definition, an AWS must meet each of the following elements:

A Weapons System: A weapon is an object"designed or used for inflicting bodily harm or physical damage" (Oxford Reference, n.d.). A system is a set of connected things or devices that operate together (Cambridge Dictionary. n.d.). A weapons system is "a weapon comprising a collection of coordinated elements, each of them essential to the system, which will not work unless all such elements are present and fully effective" (Verri and Markee, 1992).

Operates through AI: The system develops its tasks by using AI. AI is a computer program capable of problem-solving and decision-making in situations where information is incomplete or uncertain. AI embraces a broad spectrum: from previously programmed rules, which the program chooses based on inference or evaluation of options, to high-tech programs that operate through machine learning and create their scoring formula (Bathaee, 2018, p. 898).[26]

Discretionarily: AWS have a degree of freedom (Scharre, 2018, p. 53) within formerly defined guidelines to opt for a course of action. Based on previously established goals and constraints, once initiated, AWS decide on matters such as when, how, and which targets to attack and engage those targets. This means AWS are "predictable only within the range that they were programmed. Deciding which targets to engage, as well as how and when to carry out an attack, would be left to the AWS's software that is programmed to deal with a myriad of situations and changing set of circumstances" (Petman, 2017, p. 70). Discretion also explains that AWS are not random, as predefined guidelines exist. Therefore, this concept excludes devices such as land mines, which do not select targets but randomly 'attack' anything that exerts pressure.

Select targets: Select means "to choose somebody/something from a group of people or things, usually according to a system" (Oxford Learner's Dictionaries,

[26] According to Bathaee, "Artificial intelligence refers to a class of computer programs designed to solve problems requiring inferential reasoning, decision-making based on incomplete or uncertain information, classification, optimization, and perception. AI programs encompass a broad range of computer programs that exhibit varying degrees of autonomy, intelligence, and dynamic ability to solve problems. On the most inflexible end of the spectrum are AI that make decisions based on preprogrammed rules from which they make inferences or evaluate options. For example, a chess program that evaluates every possible move and then selects the best move according to a scoring formula would fall within this category. On the most flexible end are modern AI programs that are based on machine-learning algorithms that can learn from data. Such AI would, in contrast to the rule-based AI, examine countless other chess games and dynamically find patterns that it then uses to make moves – it would come up with its own scoring formula. For this sort of AI, there are no pre-programmed rules about how to solve the problem at hand, but rather only rules about how to learn from data" (Bathaee, 2018, p. 898).

2022a). A target is "an object, a person or a place that people aim at when attacking" (Oxford Learner's Dictionaries, 2022a).

Engage targets: To attack, to begin fighting against the target.

Without further intervention by a human: Once activated by a human, AWS can pursue the OODA loop without the need for human input. This means operating independently (Petman, 2017, p. 70).

We know broad concepts have the inherent problem of placing different AWS on an equal footing (Amoroso, 2020, p. 22). Nonetheless, this more comprehensive concept of AWS is a better alternative as it encompasses the whole AI spectrum, from simple targeting algorithms to future advanced AI. It provides an adequate framework for a timely discussion of AWS's responsibility challenges while not excluding future technological advancements. The provided concept has the advantage of stressing AWS's essential feature of freedom. We did not use the word autonomous to avoid tautology. The proposed concept does not coincide but is aligned with the definition of the world's leading military power, the US, the NGO coalition that advocates for an AWS ban, Stop Killer Robots, and the ICRC. Furthermore, this concept excludes automatic weapons that might perform tasks without human intervention once activated but do so randomly due to physical triggers, such as land mines, and not discretionarily and grounded on AI. We emphasize that the provided concept is not necessary for regulation but offers a framework for the weapons under discussion in this book.

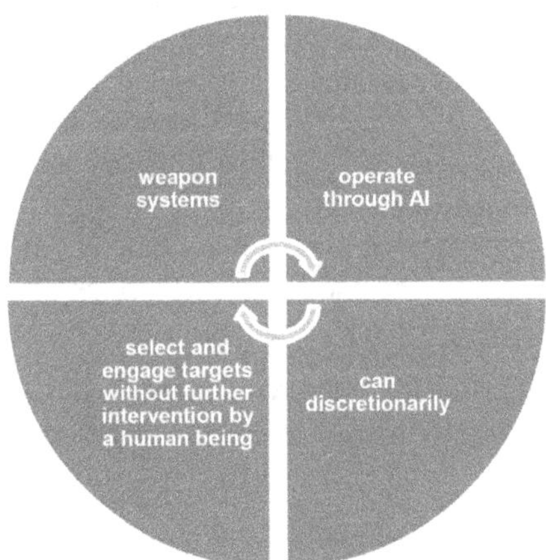

1.2 The Inherent Unpredictability of AWS

AWS are weapons that perform tasks using AI.[27] To better evaluate AWS's challenges, it is necessary to understand AWS's (species) and Autonomous Devices (genus) and their certain degree of inherent unpredictability.

[27] For AI definition see footnote 42 (Bathaee, 2018, p. 898).

The major paradigm shift autonomous systems bring about is that the problem-solving procedure is taken by machines, through AI mechanisms, and not by humans, even though humans built, activated, and set their objectives. Autonomous Devices, as the name implies, have a high degree of autonomy (IEEE, 2021a, p. 12). Therefore, they perform tasks discretionarily. In other words, all autonomous systems present a certain degree of unpredictability.

In a broad sense, predictability is the capability to "say or estimate that a specified thing will happen in the future or will be a consequence of something" (ICRC, 2019b, p. 10). In a narrow sense, predictability means "(…) knowing the process by which the system functions and carries out a specific task or function" (ICRC, 2019b, p. 10). As will be further explained, AWS's developers can only aspire to predictability in a narrow sense. Still, predictability can be challenging considering, for instance, the possibility of machine learning (Amoroso, 2020, p. 230).

The ICRC urged that unpredictable AWS shall be ruled out (ICRC, 2021).[28] Nonetheless, the ICRC also recognized that "autonomous weapon systems all introduce a degree of unpredictability into the consequences of the attack(s)" (ICRC, 2019a, p. 5).

In the following topics, we will discuss autonomous devices unpredictability (genus), and next AWS unpredictability (species).

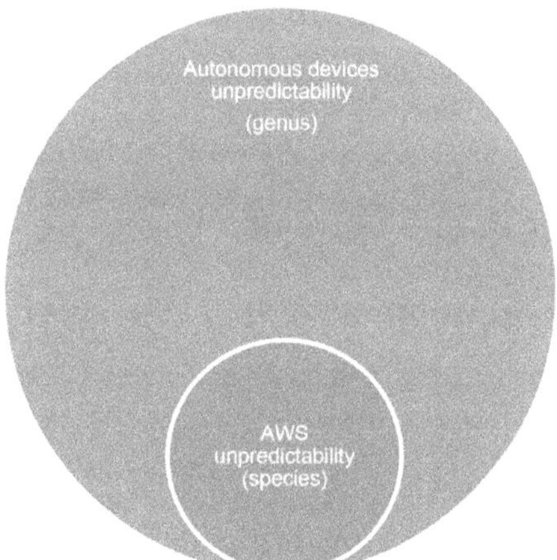

1.2.1 Autonomous Devices' Inherent Unpredictability (Genus)

The main characteristics of Autonomous Devices are under the umbrella of inherent unpredictability: task autonomy and lack of predictability (Abbott, 2020, p. 33).

[28] ICRC stated at CCW GGE that "(…) unpredictable autonomous weapon systems should be expressly ruled out, notably because of their indiscriminate effects. This would best be achieved with a prohibition on autonomous weapon systems that are designed or used in a manner such that their effects cannot be sufficiently understood, predicted and explained" (ICRC, 2021).

Task autonomy means the functional capability to determine how to accomplish a task without external support (Abbott, 2020, p. 34). The difference between task autonomy, applicable to AI, and personal autonomy, which is the illuminist framework of autonomy grounded on free will, is discussed in Chapter 4.

The lack of predictability means that AI's course of action, outcomes, and side effects might not be foreseen even to the programmers (Abbott, 2020, p. 33).

The report from the Commission to the European Parliament, which accompanied the European Union White Paper on AI, exposes features of AI that lead to unpredictability and challenge the understanding of the causes of damages. "They can combine connectivity, autonomy, and data dependency to perform tasks with little or no human control or supervision" (Commission to The European Parliament, 2020). Furthermore, AI devices are complex due to the "(…) plurality of economic operators involved in the supply chain and the multiplicity of components, parts, software, systems or services, which together form the new technological ecosystems" (Commission to The European Parliament, 2020). They are also capable of learning from experience, updates, and upgrades. In short, "The vast amounts of data involved, the reliance on algorithms and the opacity of AI decision-making, make it more difficult to predict the behavior of an AI-enabled product and to understand the potential causes of a damage" (Commission to The European Parliament, 2020).

AI systems are also vulnerable to adversarial tricking or spoofing, increasing the problem of unpredictability (ICRC, 2019a, p. 3).

In summary, the inherent unpredictability of AI in general comes from the inability to predict what paths an autonomous system will take to achieve its goals, even if the final objectives are known.[29] For instance, the developers of Deep Blue, an AI chess player, cannot foresee every turn's move but only know that its actions aim at winning (Yampolskiy, 2020, p. 110).

1.2.2 AWS Inherent Unpredictability (Species)

As AWS are a species of the autonomous devices' genre, they bear all the autonomous devices features described above. Therefore, AWS pursue a decision-making process that is likely to be different from what was anticipated by the programmers. "The more the system is autonomous then the more it has the capacity to make choices other than those predicted or encouraged by its programmers" (Sparrow, 2007, p. 70). Furthermore, AWS might learn from

[29] Yampolskiy states that "Unpredictability of AI, one of many impossibility results in AI Safety, also known as Unknowability [Vinge, 1993] or Cognitive Uncontainability [Cognitive Uncontainability, 287019], is deemed as our inability to precisely and consistently predict what specific actions an intelligent system will take to achieve its objectives, even if we know the terminal goals of the system. It is related but is not the same as unexplainability and incomprehensibility of AI [Yampolskiy, 2019]. Unpredictability does not imply that better-than-random statistical analysis is impossible; it simply points out a general limitation on how well such efforts can perform, and is particularly pronounced with advanced generally intelligent systems (super intelligence) in novel domains" (Yampolskiy, 2020, p. 110).

experience and interact with the surrounding environment, thus acting according to their learnings in the same measure or even more than their initial program. Therefore, AWS's actions will be outside the sphere of predictability and control of programmers and designers at some stage.

Moreover, the pace of AWS's decision-making processes might far surpass humans meaning thatany potential human consultation would be too late, which might enhance the unpredictability. Swarms of AWS are also a source of unpredictability, as the interaction of various devices increases complexity and thus the chances of "emergent" unforeseeable actions (ICRC, 2019a, p. 12). This topic will be discussed in detail in Part II (Chapter 5). Another factor of AI devices' unpredictability is the interaction with the environment, as the systems might perform in unanticipated ways depending on the situation on the ground. Specifically, in the case of AWS, this adds a thick layer of unpredictability as armed conflict scenarios are surprise-seeking and clumsy; thus, the environment is constantly changing.

In short, one of the critical AWS features is task autonomy, and at least a certain degree of lack of predictability, which means they are inherently unpredictable.

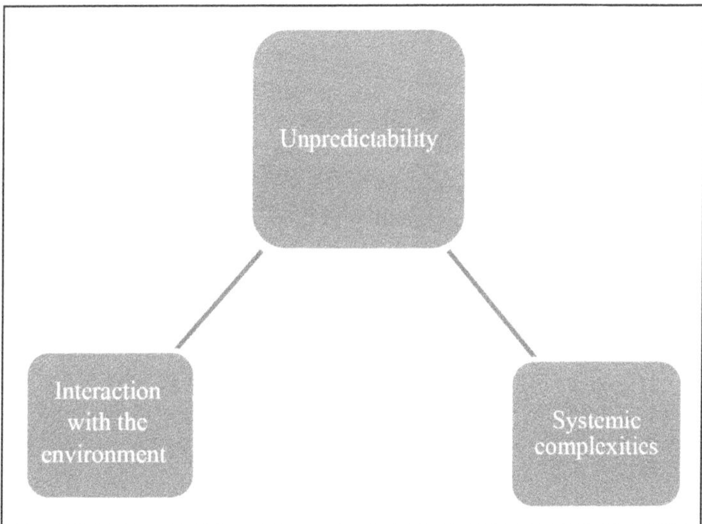

There are two main reasons for AWS`s inherent unpredictability: the autonomous systems themselves and their interactions with the environment. Therefore, unpredictability can occur throughout a broad spectrum. At one extreme, are simpler AWS,[30] such as loitering munitions like the Israeli Harpy Drone, in which the interaction with the environment is likely to be the

[30] ICRC, 2019a, recalls that "Even deterministic systems do not function in a broadly predictable fashion, owing to complexity (in design and task) and interaction with a varying environment" (ICRC, 2019a, p. 12).

predominant reason for unpredictability. At the other, there are more complex AI in which unpredictability arises due to systemic complexities, such as "black boxes," which means that it is impossible to foresee or even to trace back the decision-making process in many situations (Bathaee, 2018, pp. 891–894), as well as the system's interactions with the environment. The following sections discuss unpredictability due to systemic reasons, such as black boxes, and due to interaction with the environment.

1.2.3 Systemic Complexities: the Black Box Dilemma

We will briefly analyze some AWS unpredictability, referred to in the literature as opaqueness or the black box dilemma. Despite being programmed to make decisions based on predefined parameters, AWS actions will still be unpredictable in various cases. Unpredictability is not the same as randomness. "Mere randomness provides no support for a claim to autonomy. Instead, the actions of these machines will be based on reasons, but these reasons will be responsive to the internal states – 'desires,' 'beliefs' and 'values' – of the system itself" (Sparrow, 2007, p. 65). Significantly, autonomous systems might revise these internal states and learn from experience. "In practice, this is likely to mean that the actions of these machines will quickly become somewhat unpredictable" (Sparrow, 2007, p. 65).

The black box dilemma, or opaqueness, is, thus, the impossibility of completely comprehending an AI device's decision-making process and predicting its actions (Bathaee, 2018, p. 905). It is directly linked with AI features of lack of predictability, autonomy, and limited explainability. Limited explainability means that AI's causal chain might be inaccessible even to programmers. In short, it might be impossible to know and trace back why and how AI pursued the decision-making process (Abbott, 2020, p. 33)[31]. The lack of explainability of the outcomes implies serious difficulties in testing and assessing the performance of AWS (ICRC, 2019a, p. 2). Furthermore, AI increasingly encompasses machine learning algorithms that learn from BigData and thus arrive at dynamic solutions, which can be black boxes even for the programmers who created them (Bathaee, 2018, p. 891).

The devices' cognitive process is widely different from human beings, which is an additional obstacle to anticipating, for instance, how an AI device will identify and classify objects (Dickinson, 2018, pp. 13–14).[32] "More fundamentally, machines do not see like humans. They have no understanding of meaning or context, which means they make mistakes that a human never would" (ICRC,

[31] In this sense, Abbott affirms that "An AI's action cannot always be explained. It may be possible to determine what an AI has done, but not how or why it acted as it did" (Abbott, 2020, p. 33).

[32] Dickinson explains that "Because such a machine's internal cognitive processes are vastly different from a human's," this "significantly complicates a human's ability to predict how [the entity] might classify objects" (Dickinson, 2018, pp. 13–14).

2019a, p. 2). Regarding object recognition, for example, when machine learning processes are used, "(...) it may not be possible for a person to understand or interrogate the specific characteristics or features of the object that are providing the triggers for recognition" (Moyes, 2019, p. 7). The same occurs with facial recognition. Consequently, it will be challenging to comprehend "(...) what things might fall within the target profile but which are not intended objects of attack" (Moyes, 2019, p. 7).

Black boxes happen both in more advanced AI systems, such as deep neural networks,[33] and in less complex algorithms, such as support vector machines (Bathaee, 2018, p. 897).[34]

In deep neural networks, artificial neurons are organized in various layers that mimic human brain activity and receive and process information in a complex and non-linear venue (Montavon, Samek, and Müller, 2018, p. 2).[35] The complexity of the numerous neuron interactions leads to it being practically impossible to comprehend the decision-making process (Bathaee, 2018, p. 897).[36] They can be as challenging to understand as the human brain. The unpredictability arises from the complexity of the AI's algorithm (Bathaee, 2018, p. 901).

A support vector machine is an algorithm that learns how to label objects from examples (Techopedia, 2016; Noble, 2006). They perform nonbinary classification in a higher-dimensional space (Adankon and Cheriet, 2009). Despite being less complex than deep neural networks, the support vector machine's decision-making process is also a black box for programmers and operators due to its high dimensionality. Those algorithms can find geometric patterns in

---

33 Techopedia explains that "A neural network, in general, is a technology built to simulate the activity of the human brain – specifically, pattern recognition and the passage of input through various layers of simulated neural connections.
Many experts define deep neural networks as networks that have an input layer, an output layer, and at least one hidden layer in between. Each layer performs specific types of sorting and ordering in a process that some refer to as "feature hierarchy." One of the key uses of these sophisticated neural networks is dealing with unlabeled or unstructured data. The phrase "deep learning" is also used to describe these deep neural networks, as deep learning represents a specific form of machine learning where technologies using aspects of artificial intelligence seek to classify and order information in ways that go beyond simple input/output protocols" (Techopedia, 2023).

34 Techopedia explains that "A support vector machine is a supervised learning algorithm that sorts data into two categories. It is trained with a series of data already classified into two categories, building the model as it is initially trained. The task of an SVM algorithm is to determine which category a new data point belongs in. This makes SVM a kind of non-binary linear classifier" (Techopedia, 2016).

35 Montavon, Samek, and Müller explain that a Deep Neural Network "is a collection of neurons organized in a sequence of multiple layers, where neurons receive as input the neuron activations from the previous layer, and perform a simple computation (e.g. a weighted sum of the input followed by a nonlinear activation). The neurons of the network jointly implement a complex nonlinear mapping from the input to the output" (Montavon, Samek, and Müller, 2018, p. 2).

36 Bathaee affirms that "(...) deep neural network, which often involves the use of thousands of artificial neurons to learn from and process data. The complexity of these countless neurons and their interconnections makes it difficult, if not impossible, to determine precisely how decisions or predictions are being made" (Bathaee, 2018, p. 897).

higher-dimensional space, which humans cannot visualize. Therefore, it may not be possible to truly understand how a trained AI program arrives at its decisions or predictions (Bathaee, 2018, pp. 891–894).

The more complex the system embracing multiple sensors that collect data and highly complex algorithms to evaluate such data is, the less predictable the output and the process to arrive at it will be (ICRC, 2019a, p. 12). Thus, the causation chain will not be transparent or even exist at all.

Black boxes can be labeled as strong and weak. They are strong when there is a total inability to predict and comprehend how an AI decides on a course of action. They are weak when, despite the difficulties, there is still a limited possibility of predicting actions and understanding the decision-making process (Bathaee, 2018, p. 901). In both cases, the decision-making process is, to some extent, opaque to humans (Ali, 2020, pp. 16–17). Thus, AI devices do not follow pre-written human instructions but make decisions based on data and possibly adapting algorithms.

1.2.4 Unpredictability Due to Interaction with the Environment

Unpredictability might depend not only on AWS's features but also on the complexity of the task, the kind of environment, and their interactions with the environment to accomplish the task (ICRC, 2019a, p. 11). With autonomous devices, the control is progressively reallocated from the designer to the device itself. In the same proportion that the creator's control over the AI device reduces, the impact of the operational environment increases. Thus, autonomous devices' behavior cannot be predicted on previous fixed rules since actions are shaped according to the environmental inputs (Matthias 2004, 182),[37] creating inevitable "accountability gaps" (Matthias 2004, 182; Amoroso 2020, 123,124; Heyns, 2014, p. 46).[38]

---

[37] Matthias explains that an autonomous agent "(…) acts outside the observation horizon of its creator, who, in the case of a fault, might be unable to intervene manually (because he might not know about the fault until a much later point in time). This is the case for pure information agents (internet indexing programs) as well as for agents which have a physical manifestation (autonomous space vehicles, targeting systems of military missiles, mobile electronic pets). Thus, we can identify a process in which the designer of a machine increasingly loses control over it, and gradually transfers this control to the machine itself. In a steady progression the programmer role changes from coder to creator of software organisms. In the same degree as the influence of the creator over the machine decreases, the influence of the operating environment increases. Essentially, the programmer transfers part of his control over the product to the environment. This is particularly true for machines that continue to learn and adapt in their final operating environment. Since in this situation they have to interact with a potentially great number of people (users) and situations, it will typically not be possible to predict or control the influence of the operating environment" (Matthias 2004, 182).

[38] Heyns, for example, highlights accountability challenges due to interactions with the environmental unpredictability by stating that "The argument would be that the machine took its own decisions, which are unpredictable, not because computers act randomly, but because the environments in which they operate are so complex that all possible interactions between the system and the surroundings cannot be foreseen. Even if accountability can in theory be assigned by law, in practice those who activate autonomous weapon systems may find a lot of sympathy

(Contd.)

Significantly, "An autonomous robotic system that functions predictably in a specific environment may become unpredictable if that environment changes or if it is used in a different environment" (ICRC, 2019a, p. 11). AI machines perform well in developing one task, and so far, they are not as good as humans at understanding the broader context (Scharre, 2016; Scharre 2018, p. 232). Furthermore, law enforcement and armed conflict scenarios are constantly changing, unorderly, and surprise-seeking (Tamburrini, 2016, p. 139; Ford, 2017, p. 436; Scharre, 2016). Thus, all sorts of interactions with the environment are impossible to predict. Unless AWS operate in the deep sea or the desert, the interaction machine environment is a significant source of unpredictability (Ford, 2017, pp. 437–438; Marino and Tamburrini, 2006, p. 47).[39] The more complex the system and the environment, the more difficult it is to predict its courses of action and its interactions with the environment (Scharre, 2016). Even programmers might be unable to predict possible outcomes (Scharre, 2018, p. 36; Marino and Tamburrini, 2006). When AWS have self-learning and self-reprograming feature, they might diverge from the instructions humans initially installed and change their target profiles based on the operational environment (Moyes, 2019, p. 7).

AWS's unpredictability is further complicated as AWS belonging to different parties might interact in a conflict. "Therefore, it could be plausibly maintained that, no matter how they are designed, autonomous weapons systems are bound to act in ways that cannot be foreseen in advance by their users in a non-negligible number of cases" (Amoroso, 2020, pp. 126–127).

The Brazilian delegation has stressed at the CCW the unpredictability considering that the "environment that informs the operator is mathematically modeled and may be subject to misunderstandings and malfunctions; human errors may thus be replaced by cyber misinterpretations of the environment or situational awareness, or by system biases" (Brazilian Delegation, 2020a).

In summary, AWS are inherently unpredictable since their interactions with the environment might be unexpected. Simpler or more complex algorithms might be black boxes even to their creators, and so far, AI is task-specific and incapable of thoroughly understanding a broader context. Swarming and spoofing also increase unpredictability.

---

from judges and others who have to assess their conduct. The danger of an accountability gap, in law or in practice, remains" (Heyns, 2014, p. 46).

39 Ford affirms that "The interaction between environment and machine is critical, as the ability of a system to distinguish is a function of the sophistication of the system and the complexity of the environment. An increasingly complex environment requires an increasingly sophisticated system" (Ford, 2017, pp. 437–438).

In the same sense, Marino and Tamburrini affirm that "(…) an epistemological reflection on computational learning theories and machine learning methods suggests that programmers and manufacturers of learning robots may not be in the position to predict exactly what these machines will actually do in their intended operation environments" (Marino and Tamburrini, 2006, p. 47).

CHAPTER

2

The Accountability Gap, the Focus on Individual Criminal Responsibility, and the Shadow of State Responsibility until 2022

International law, both customary and conventional (UNGA, 1998, preamble; Convention on the Prevention and Punishment of the Crime of Genocide, 1948, art. 6; Convention Against Torture and Other Cruel, Inhuman or Degrading Treatment or Punishment, 1984, art. 7; Protocol I, 1978, art. 86), requires that States hold accountable those who commit breaches, which includes individual responsibility and State responsibility (Ford, 2017, p. 462).

The obligation to hold accountable is a corollary of the international rule of law (UN Secretary-General, 2004, p. 4).

> From key provisions of the Geneva Conventions and additional protocols, it flows that states must be able to discern, scrutinize and attribute conduct that is a potential breach of IHL (…). In the case of a grave breach—that is, a serious violation of IHL that amounts to a war crime—states are obliged to initiate investigations, 'search for' individuals who have committed (or otherwise facilitated) a breach, and 'repress' such conduct. In the case of any other IHL violation (not amounting to a grave breach), states are obliged to 'suppress' such violations. These obligations are reflected in IHL as well as in customary international law, and failing to conduct investigations of allegations of serious violations of IHL, including war crimes, constitutes non-compliance with a state's international obligations (Bo, Bruun and Boulanin, 2022).

Nonetheless, actions perpetrated in the context of AWS might challenge existing legal norms and create additional obstacles to responsibility.

Furthermore, while IHL requires those who commit breaches to be held accountable, IHL does not require individual responsibility every time the use of force decision is taken, but only that States are responsible for their actions (Scharre, personal communication, May 27, 2022).[1]

In the context of AWS, there might exist a vacuum of individual responsibility due to the autonomous nature of the system, but if a violation occurs, States need to be held responsible (Garcia, personal communication, June 13, 2022).[2]

Most of the literature is devoted to individual responsibility, and the State's responsibility has remained in the shadows, as discussed throughout this chapter. However, State responsibility plays a key role when breaches of IHL occur.

In the following sections, we will briefly present the debates over personal responsibility and highlight how State responsibility has been in the shadows, at least until 2022, to set the ground for the research topic: State responsibility, which will be addressed in depth in Part II.

2.1 AWS and the Individual Accountability Gap

The AWS' individual accountability gap debate navigates through a spectrum with two extremes. Those who struggle to ascribe responsibility are on one end. On the other end are those who state that ascribing responsibility for

---

1 This is a citation from the author's Ph.D. dissertation. During an interview for the author's Ph.D., Paul Scharre, in his personal capacity, stated that: "a lot of the debate around an accountability gap is centered on individual accountability. And it's important to keep in mind that the law of war doesn't require individual accountability for use of force decisions. It does not require today that any time there's a decision to use force there is some individual that we can hold up as a scapegoat if something goes wrong. What is required is that the State is responsible, but the law of war does not mandate individual accountability. If individuals commit war crimes, they should be held accountable, but there are situations where today responsibility and decision-making for the use of lethal force is distributed among many individuals, situations where some individuals are assessing targets and are part of an approval process to recommend, and then approve a targeting list. Then others approve the launching of strikes based on this targeting list. And then others actually carry out the strikes. The decision-making is distributed among a variety of actors."

2 This is a citation from the author's Ph.D. dissertation. During an interview for the author's Ph.D., Eugênio Garcia, in his personal capacity, stated that:
"The problem is that if these weapons are autonomous, they are deciding for themselves, it's like there's a vacuum. The gap exists due to the nature of the system that is autonomous, but you need to ensure accountability. There's no way you can claim: 'no one is responsible because it's an autonomous weapon, you can blame the weapon, period. IHL is aimed at defending the person, the human being, and protected places. If people or protected places were hit, someone has to be held accountable. It's no excuse that the gun is autonomous so that no one will be held accountable, that everyone will wash their hands, that doesn't work. There's a violation, someone has to be held accountable. In the case of IHL it will be the State." My translation of the original in Portuguese "O problema é que se estas armas são autônomas, elas estão decidindo por si mesmas, é como se tivesse um vácuo. O gap existe pela natureza do sistema que é autônomo, mas você precisa ter responsabilização. Não tem como você alegar: 'ninguém é responsável porque é uma arma autônoma, você pode culpar a arma, paciência'. O DIH é voltado para a defesa da pessoa, do ser humano, e de lugares protegidos. Se pessoas ou lugares protegidos foram atingidos, alguém tem que ser responsabilizado. Não é desculpa que a arma seja autônoma para que ninguém será responsabilizado, que todo mundo vai lavar as mãos, isto não funciona. Há uma violação, alguém tem que ser responsabilizado. No caso do DIH será o Estado."

AWS misdoings is bound to fail (Amoroso, 2020, p. 122). There are mainly two reasons for the accountability challenges: The above-mentioned inherent unpredictability of AWS and the "many hands" problem. This last means that many people are involved in the chain of action, so none of them can be deemed responsible for the breach (Amoroso, 2020, p. 123).

The unpredictability of AWS is a significant cause of the accountability gap, considering "(…) varying degrees of uncertainty as to exactly when, where and/or why a resulting attack will take place" (ICRC, 2018, p. 2). Predictability is necessary to connect human behavior and intent (*mens rea*) to AWS misdoings (ICRC, 2018, p. 2). Thus, no human being might be deemed responsible. Also, if autonomous devices are not explainable and function as black boxes, the likelihood of foreseeing AWS' actions decreases or vanishes,[3] and it is challenging to comprehend the root causes of a system's behavior which might impede ascribing responsibility (Garcia, personal communication, June 13, 2022).[4]

As early as 2007, Sparrow, who sparked the debate on AWS and the problem of the accountability gap (Amoroso, 2020, pp. 121–122), highlighted that "A

3 In this sense, The US Defense Advanced Research Projects Agency (DARPA) is trying to develop new explainable AI projects called eXplainable AI, or XAI. As stated by the US Delegation at CCW GGE, "trust and accountability issues are posed by the fact that current AI systems often use processes that are opaque to the human operators of the systems. To help address trust and accountability issues, the Defense Advanced Research Projects Agency's Explainable AI project seeks to develop new machine-learning systems that "have the ability to explain their rationale, characterize their strengths and weaknesses, and convey an understanding of how they will behave in the future." By seeking to develop AI systems that are more transparent to human operators, such work in the research and development area can address concerns that might be posed using such technology."(US Delegation, 2018). Amoroson comments that "Pending significant breakthroughs in XAI, one cannot but acknowledge the present technological difficulty of ensuring sufficient levels of system interpretability and explainability (…)" (Amoroso, 2020, p. 248).

Bathaee, dealing with personal responsibility and the gap of intent and causation regarding AI, and not with AWS and State responsibility, states that two possible paths are regulating the level of transparency required from AI devices and establishing strict liability. He further acknowledges that both ways are inadequate. Transparency would hinder technological development and there is also no certainty as to if AI devices can be "(…)developed with increased transparency. The future may in fact bring even more complexity and therefore less transparency in AI, turning the transparency regulation into a functional prohibition on certain classes of AI that inherently lack transparency" (Bathaee, 2018).

4 This is a citation from the author's Ph.D. dissertation. During an interview for the author's Ph.D., Eugênio Garcia, in his personal capacity, stated that: "You're going to apply old rules to a certain extent, you have to be clear and separate what can be regulated from what's impossible to regulate, because sometimes you won't be able to assign responsibility if you don't understand how that decision was reached, it's the famous black box, where you put the data and then you have a result, but you don't know how the machine came to this conclusion" – and underscored the Red Cross's position that unpredictable systems should be banned." My translation from the original in Portuguese "Você vai aplicar regras antigas até um certo ponto, você tem que ter uma clareza e separar o que pode ser regulado do que é impossível regular, porque às vezes você não vai ter como atribuir responsabilidade se você não compreende como se chegou àquela decisão, é a famosa caixa preta, que você coloca os dados e depois tem um resultado, mas não sabe como é que a máquina chegou a esta conclusão." – e ressaltou a posição da Cruz Vermelha de que sistemas imprevisíveis devem ser proibidos.

number of possible *loci* of responsibility for robot war crimes are canvassed: the persons who designed or programmed the system, the commanding officer who ordered its use, the machine itself" (Sparrow, 2007, p. 62). However, "none of these are ultimately satisfactory" (Sparrow, 2007, p. 62). From an ethical standpoint, he also raises the necessity to hold someone accountable for IHL violations and considers it unethical to deploy AWS due to the personal accountability gap (Sparrow, 2007).

AWS challenges the paradigm that has prevailed since the Nuremberg trials, whereby international crimes are only committed by man and enforcing international law entails individual punishment (Dickinson, 2018, pp. 4–5; Crootof, 2016, pp. 1385–1386).[5] There might be breaches of international law in the context of AWS, which, if committed by humans, would be a war crime. Paradoxically, as they were performed in the context of AWS, no human being might be held responsible due to the lack of intentionality, neither the programmer, developer, manufacturer, or deployer (Crootof, 2016, pp. 1385–1386; Scharre, personal communication, May 27, 2022).[6]

Accountability gaps may also happen with using other weapons when accidents occur (if there is no intentionality). AWS widens this gap due to the shift from humans in the decision-making paradigm to AWS which can independently select and engage targets. Therefore, humans might be unable to predict the paths AWS will follow. International criminal responsibility requires intent and knowledge (UNGA, 1998, art. 30), which means acting willfully or recklessly (Crootof, 2018, pp. 64–67).

---

5 Dickinson highlights that "(…) autonomy poses serious accountability challenges. In particular, these systems threaten the framework of individual criminal responsibility that many would argue undergirds all LOAC/IHL. From the Nuremberg trials of Nazi war criminals, to proceedings before more recently established international courts and tribunals, to domestic civilian and military prosecutions, a fundamental principle of LOAC/IHL is that human beings will be held individually responsible for egregious violations of the law of war that constitute war crimes. While such prosecutions are rare, they constitute one of the core sanctions that seek to ensure compliance with this body of law. Yet autonomous weapons pose problems for this framework. As a number of scholars have pointed out, in the case of truly autonomous systems, who will be responsible for the decision to strike? In many instances there may be no human being with the requisite level of intent to trigger individual responsibility under existing doctrine" (Dickinson, 2018, pp. 4–5). In the same sense Rebecca Crootof (Crootof, 2016, pp. 1385–1386).

6 Crootof argues that "Autonomous weapon systems challenge a presumption that undergirds all of international criminal law: that serious violations of international humanitarian law will not occur absent willful human action. In situations where no one acts intentionally or recklessly, under current law no one—not the deployer, commander, programmer, developer, manufacturer, or the weapon system itself—can (or should) be held criminally liable for the deadly consequences of an autonomous weapon systems' unanticipated actions" (Crootof, 2016, pp. 1385–1386). This is a citation from the author's Ph.D. dissertation. During an interview for the author's Ph.D., Paul Scharre, in his personal capacity, stated that: "that is the sort of uncomfortable space where there is an accountability gap. That you can have an AWS committing an act that if committed by a person would be a war crime, but is not a war crime because there is no intentionality, because the person who launched the weapon or the person who programmed it didn't know it was going to do that. While an accountability gap might exist in some circumstances, it is not unique to AWS. (…) Accidents happen in daily life (...) Do AWS widen that gap? maybe that is a valid question, but it is not a wholly novel problem."

> Unlike conventional weapons or remotely operated drones, autonomous weapon systems can independently select and engage targets. As a result, they may take actions that look like war crimes—the sinking of a cruise ship, the destruction of a village, the downing of a passenger jet—without any individual acting intentionally or recklessly. Absent such willful action, no one can be held criminally liable under existing international law (Crootof, 2016, p. 1348).

International Humanitarian Law has focused on individual criminal responsibility to enforce the law. However, with AWS, a decision is so fragmented that none of those in the chain of action might have sufficient knowledge and intent to be held responsible (Dickinson, 2018). If a person intentionally deploys AWS to commit violations, they will be held criminally responsible. Nonetheless, in the context of AWS, they pursue the OODA loop autonomously. Due to the unpredictability of the algorithm itself or the interaction of AWS with the environment,it may be that there is no individual who can be held criminally liable, unless that person intentionally or recklessly deployed AWS to breach the law (Crootof, 2018, pp. 64–67).

Those who advocate for a treaty prohibiting the use of AWS have, among other arguments, claimed that AWS generate insurmountable barriers to responsibility (Wareham, 2019).[7]

> "(…) the use of fully autonomous weapons would lead to a gap in individual criminal responsibility for war crimes. Commanders are responsible for the actions of a subordinate if they knew or should have known the actions would be unlawful and did not prevent or punish them. It would be legally challenging, and unfair, to hold commanders liable for the unforeseeable actions of a machine operating outside their control" (Docherty, 2019a).

The HRW report "Mind the Gap–The Lack of Accountability for Killer Robots" highlighted many roadblocks to accountability for AWS-related actions. Criminal liability would be applicable only when there was a clear human intent to commit breaches. Moreover, civil liability,for instance, in the US is challenging due to the immunity of the military and its contractors (Docherty 2015, p. 37). Regardless of the jurisdiction, AWS lawsuits can be complex and expensive since they require legal and technical experts, which might deter litigation in the domestic field (Docherty, 2015, p. 27).

Liability gaps are more severe in the military field due to three factors: the principle of military necessity that justifies the deployment of AWS; States have no interest in unilaterally restricting harmful technologies that put them in a situation of strategic advantage; and the restricted scope for civil liability in the absence of a human criminally responsible for the breach (Sartor and Omcini, 2016, pp. 70–71). AWS might perform tasks in venues that "(…) (a) they would

---

7 Note that McDougall has criticized such approach of advocating for a ban on the ground of an accountability gap (McDougall, 2019, p. 58).

represent material violations of criminal laws of war, (b) they would not lead to any proven responsibility of human individuals, and (c) they would cause damage for which victims could not receive any remedy" (Sartor and Omcini, 2016, p. 71).

In search of responses to misdoings in the context of autonomous devices, some authors argue for expanding "criminal or civil penalties directed at people" (Abbott, 2020, p. 16). Such an approach seems problematic and has not so far found resonance at the international level that grounds responsibility on intent, causation, and foreseeability.

Despite the considerable scholarship on the accountability gaps, some argue that there is no such gap (Kalmanovitz, 2016, pp. 145–146),[8] since humans remain in the "wider loop"of the decision-making process because they program the algorithms (Lieblich and Benvenisti, 2016, p. 250).[9] This view maintains that "any commander who decides to launch AWS into a particular environment is, as with any other weapon systems, accountable under international criminal law for that decision" (Schmitt, 2015). They also claim that developers will be accountable if they design systems that conduct operations that do not comply with IHL (Schmitt, 2015). Some claim to slightly revise the doctrine of command responsibility to extend it to the conduct of AWS (Margulies, 2016, p. 25).[10]

In short: many authors have devoted efforts to the individual responsibility gap. The scholarship is divided among those who defend that commanders, designers, and manufacturers have criminal responsibility and those who identify unsurmountable challenges in assigning responsibility to those actors. Nonetheless, few have directed their attention toward State responsibility.

Some scholars claim that since autonomous devices might commit acts analogous to crimes and no natural person might be held liable,[11] the autonomous

8 Kalmanovitz, affirms that "(…) AWS can conceivably be developed and deployed in ways that are compatible with IHL and do not preclude the fair attribution of responsibility, even criminal liability, in human agents. While IHL may significantly limit the ways in which AWS can be permissibly used, IHL is flexible and conventional enough to allow for the development and deployment of AWS in some suitable accountable form (…)" (Kalmanovitz, 2016, pp. 145–146).

9 Lieblich and Benvenisti's view is that "Since a technically autonomous machine is incapable of altering its algorithms though a process that could be equated to human learning, we cannot claim that it engages in a true process of discretion – its 'inner deliberation' is controlled by determinations made ex ante by humans. Risking oversimplification – and ongoing research notwithstanding – it is safe to say that computers are generally unable to engage in 'thinking about thinking' or metacognition" (Lieblich and Benvenisti, 2016, p. 250).

10 In Margulie's perspective, "Application of command responsibility to the actions of AWS requires a modest revision of the doctrine, which would extend its application to the acts of machines. That extension is a logical outgrowth of an evolution in the conduct of hostilities, in which more warfighting is done autonomously. As circumstances on the ground change, IHL should evolve, if it is to continue to preserve the balance of military necessity and humanity" (Margulies, 2016, p. 25).

11 Abbott affirms that "(…) AI is already autonomously engaging in activities that would be criminal for a natural person. Moreover, AI can do so in a way that is untraceable, or irreducible to the wrongful act of a person. (…) there are cases of AI generated crimes where no natural person can be held criminally liable" (Abbott, 2020, p. 13).

device itself should be held criminally responsible (Abbott, 2020, p. 13).[12] Despite being ethically reprovable, criminal responsibility being assigned to autonomous devices is, from a juridical standpoint, not as absurd as it might first appear since, through legal fiction, corporations have already been held accountable (Abbott, 2020, p. 14).[13] In the context of corporations,there are decisions straightforwardly attributable to individuals,but there are also decisions taken by "many hands" and based on "corporate culture" that might lead to crimes. In the context of autonomous devices' criminal responsibility, punishment would aim at impacting the behavior of developers and users and could include the confiscation or destruction of autonomous devices and financial penalties (Abbott, 2020, p. 15).[14] There are pushbacks to punishing autonomous devices since it could send a disturbing message of morally equating autonomous devices to people (Abbott, 2020, p. 16), and AWS fall outside the jurisdiction of international tribunals (Docherty 2015, p. 37). Furthermore, it is questionable how practical this approach would be except for recognizing crime on a symbolic level. This kind of responsibility might be further discussed within the realm of business and human rights but is outside the scope of this book.

2.2 The Lack of Focus on State Responsibility

"Ironically, as the technologies of State violence become increasingly sophisticated and brutally lethal (...), international institutions have narrowed their targets of legal responsibility to a handful of individuals" (Fletcher, 2016, p. 451).

The international responsibility of States is the new juridical obligation that arises from a breach of international law. It is the necessary corollary of

---

12 Abbott claims that "A small but growing number of academics are advancing such arguments, and criminal punishment for AI might seem to follow from the principle of AI legal neutrality, which cautions against different legal treatment of AI and human behavior" (Abbott, 2020, p. 13).

13 Abbott argues that "(...) criminal punishment of AI is not as ridiculous as it may first appear. The law already criminally punishes artificial persons in the form of corporations. Even though they do not literally possess mental states, corporations can directly face charges when their defective procedures cause harm, particularly where structural problems in corporate systems and processes are difficult to reduce the wrongful actions of individuals. The law criminally punishes strict liability offenses, acts not requiring any wrongful mental state such as intent to cause harm. Punishment can even be imposed on failure to act. In sum, punishing an artificial person for failing to act, even without evidence of harmful intent, is not something that can be dismissed out of hand. Criminal law can – and, where corporations are involved, already does – appeal to elaborate legal fictions to provide a basis for punishing some artificial entities" (Abbott, 2020, p. 14).

14 Abbott explains that "The purpose would be to impact the behavior of AI developers, owners and users. This could occur if punishment were to involve confiscation or destruction of a valuable AI or financial penalties directed at AI owners. AI punishment could also psychologically benefit victims of AI-generated crimes who would see the state affirm their rights and punish the entity that caused them harm. It would reassure citizens that antisocial activity, even by AI, will not be tolerated" (Abbott, 2020, p. 15).

international law (ICJ, 1970, par. 36), a central lynch pin of international society since it conveys the binding nature of international obligations (Geiß, 2015, p. 21). States are responsible for breaches of international law committed by their agents and war crimes perpetrated by their armed forces. States might even be responsible for *ultra vires* acts committed by their agents and crimes committed by non-State actors in situations such as delegation of power, direction, and control (Crootof, 2016, pp. 1357–1358).

In the aftermath of World War II,especially after the Nuremberg trials,[15] State responsibility for breaches of IHL has been somewhat eclipsed by the rise of individual criminal responsibility (Crootof, 2016, p. 1354).[16] The language shifted from breaches of international humanitarian law to “war crimes” (Crootof, 2016, p. 1386), which contributes to the lessening of State responsibility, “implying that there cannot be a serious violation without a morally culpable perpetrator” (Crootof, 2016, p. 1386). The international community’s focus has been more on regimes that hold individuals responsible for crimes committed on behalf of or using the structure of States or that they should have prevented.

> “If part of the rationale for individual accountability is that no one should be able to commit egregious international harms without consequence, why do we tolerate a different standard for States? It is curious that international criminal law has assumed the moral apex of international condemnation for mass atrocities to the exclusion of punishing States for the same conduct” (Fletcher, 2016, p. 453).

Even though human rights violations committed in the context of armed conflicts often trigger litigation on State responsibility, the regime of State responsibility is rarely raised as an enforcement tool for international humanitarian law as such. While institutions aiming at assuring individual criminal responsibility, such as the *ad hoc* international criminal tribunals and the International Criminal Court (ICC), developed, State responsibility for internationally wrongful acts seems to have stagnated. The leading global forum for States is the International Court of Justice (ICJ), where jurisdictional barriers and a lack of enforcement mechanisms exist (Hammond, 2015, p. 656).[17]

---

15 Fletcher explains that “(...) the Nuremberg model of accountability excludes collective forms of sanction. Remember that part of the reason the Allies decided to criminally prosecute the Nazi leaders was to decouple the Nazi regime from the German state and to demonstrate that the German people as a whole were not on trial” (Fletcher, 2016, pp. 490–491).

16 Rebecca Crootof affirms that“While the development of international criminal law is laudable, the tendency of some to treat individual criminal accountability as the sole remedy to violations of international humanitarian law is a mistake. Although certain war crimes may be committed by individuals, wars “are fought between political communities and by groups.” Focusing on individual criminal liability tends to obscure the fact that states remain legally responsible for serious violations of international humanitarian law. The law of state responsibility is rarely discussed or enforced with regard to serious violations of international humanitarian law, even though most such violations would be more appropriately attributed to the state than to individuals” (Crootof, 2016, p. 1364).

17 Hammond acknowledges that “Unfortunately, the sharp limitations on its personal jurisdiction will likely obstruct its power to hear AWS disputes. It also lacks an enforcement mechanism to

(Contd.)

Explaining the reason for the lack of focus on State responsibility is outside the scope of this book, since it aims to question if the existing international regime of international responsibility of States suffices or not, and for that, it suffices to identify the phenomenon.

We do not call for changing the course of the debate, but that the two paths of individual and State responsibility are taken as far as possible. AWS highlights this lack of focus on State responsibility, as individual responsibility might face many challenges (Crootof, 2016, p. 1366).[18] When breaches occur, the international community will call for responsibility, and State responsibility must be recognized. Furthermore, State responsibility will develop a central role when it is the only viable source of international responsibility. According to ILC commentaries, ARSIWA includes treaty breaches, torts, and crimes under the term "internationally wrongful act" and does not make any distinction between them (UN ILC, 2001b, p. 55).[19]

2.2.1 The Ghost of State Responsibility at the CCW GGE

At CCW, the main forum of discussions on AWS, the focus is also on the individual rather than State responsibility. The GGE endorsed 11 guiding principles during the 2019 Meeting of the High Contracting Parties.[20] Principles "b" and "d" deal with accountability:

provide remedies for those cases that might arise. These factors undermine the ICJ's effectiveness in terms of its ability to hold states liable for AWS crimes" (Hammond, 2015, p. 656).

It is important to recall the importance of the regional human rights courts in assuring State responsibility. See for instance Quell, stating that "The conflict in Ukraine also gave rise to a series of cases in the European Court against Ukraine and Russia (Quell, 2022).

18 Crootof explains that "The rise of individual criminal responsibility has had the unfortunate side effect of eclipsing the role of the law of state responsibility for serious violations of international humanitarian law. Institutions for holding individuals accountable for war crimes have flowered, while institutional approaches to holding states accountable for their internationally wrongful acts in armed conflicts have stagnated. Although some have expressed concern at this apparent tradeoff, up until now it has not been obviously problematic. Because they may take unpredictable action, however, autonomous weapon systems break the causal chain necessary for individual criminal liability and thereby highlight the relative lack of institutional means of holding states responsible for serious violations of international humanitarian law" (Crootof, 2016, p. 1366).

19 ARSIWA commentaries explain that "(…) there is no room in international law for a distinction, such as is drawn by some legal systems, between the regime of responsibility for breach of a treaty and for breach of some other rule, i.e. for responsibility arising *ex contractu or ex delicto*. In the "Rainbow Warrior" arbitration, the tribunal affirmed that "in the field of international law there is no distinction between contractual and tortious responsibility." As far as the origin of the obligation breached is concerned, there is a single general regime of State responsibility. Nor does any distinction exist between the "civil" and "criminal" responsibility as is the case in internal legal systems" (UN ILC, 2001b, p. 55).

20 The GGE agreed in 2018 on ten guiding principles:

"21. It was affirmed that international law, in particular the United Nations Charter and international humanitarian law (IHL) as well as relevant ethical perspectives, should guide the continued work of the Group. Noting the potential challenges posed by emerging technologies in lethal autonomous weapons systems to IHL, the following were affirmed, without prejudice to the result of future discussions:

(a) International humanitarian law continues to apply fully to all weapons systems, including the potential development and use of lethal autonomous weapons systems.

(*Contd.*)

(b) Human responsibility for decisions on the use of weapons systems must be retained since accountability cannot be transferred to machines. This should be considered across the entire life cycle of the weapons system.
(d) Accountability for developing, deploying and using any emerging weapons system in the framework of the CCW must be ensured in accordance with applicable international law, including through the operation of such systems within a responsible chain of human command and control.

As can be observed, the main focus is on human responsibility since principle "b" addresses only human responsibility, and principle "d" refers to a chain of command and control that is related to individual responsibility. None of the principles mentions State responsibility expressly.[21] While commenting on the

---

(b) Human responsibility for decisions on the use of weapons systems must be retained since accountability cannot be transferred to machines. This should be considered across the entire life cycle of the weapons system.
(c) Accountability for developing, deploying, and using any emerging weapons system in the framework of the CCW must be ensured by applicable international law, including through the operation of such systems within a responsible chain of human command and control.
(d) In accordance with States' obligations under international law, in the study, development, acquisition, or adoption of a new weapon, means or method of warfare, determination must be made whether its employment would, in some or all circumstances,be prohibited by international law.
(e) When developing or acquiring new weapons systems based on emerging technologies in the area of lethal autonomous weapons systems, physical security, appropriate non-physical safeguards (including cyber-security against hacking or data spoofing), the risk of acquisition by terrorist groups and the risk of proliferation should be considered.
(f) Risk assessments and mitigation measures should be part of the design, development, testing and deployment cycle of emerging technologies in any weapons systems.
(g) Consideration should be given to the use of emerging technologies in the area of lethal autonomous weapons systems in upholding compliance with IHL and other applicable international legal obligations.
(h) In crafting potential policy measures, emerging technologies in the area of lethal autonomous weapons systems should not be anthropomorphized.
(i) Discussions and any potential policy measures taken within the context of the CCW should not hamper progress in or access to peaceful uses of intelligent autonomous technologies.
(j) The CCW offers an appropriate framework for dealing with the issue of emerging technologies in lethal autonomous weapons systems within the context of the objectives and purposes of the Convention, which seeks to strike a balance between military necessity and humanitarian considerations" (CCW GGE, 2018).
In 2019 an additional guiding principle was agreed:
"Human-machine interaction, which may take various forms and be implemented at various stages of the life cycle of a weapon, should ensure that the potential use of weapons systems based on emerging technologies in the area of lethal autonomous weapons systems is in compliance with applicable international law, in particular" (CCW GGE, 2019a).
For deeper comprehension, see Appendix A and Appendix G.

21 Note that, at some points, GGE CCW did touch upon State responsibility but devoted only minor attention to it.
"17. On the agenda item 5 (a) "An exploration of the potential challenges posed by emerging technologies in the area of lethal autonomous weapons systems to International Humanitarian Law" the Group concluded as follows: (…)
(c) States, parties to armed conflict and individuals remain at all times responsible for adhering to their obligations under applicable international law, including IHL. States must also ensure

(Contd.)

principles, many States reaffirmed that AWS lack agency and a legal personality and claimed that a policy framework must be directed to humans (Chair of the 2020 GGE LAWS, 2020).[22]

State responsibility was seldom raised at CCW GGE until 2022. Except for the few comments we will present here, the focus remained on human responsibility (Chair of the 2020 GGE LAWS, 2020). In 2018 the Non-Aligned Movement (NAM) called for a legally binding instrument on AWS that embraces the issue of State responsibility. "(…) NAM believes that the following elements shall be included in the substantive discussion of this matter: (…) (b) The responsibility of States for internationally unlawful acts caused by lethal autonomous weapons systems" (Bolivarian Republic of Venezuela, 2020). In its 2020 statement, the US highlighted the lack of focus on State responsibility and the need for GGE to address it expressly as regards AWS (US Delegation, 2020).[23] The 2020 Joint Commentary by Austria, Belgium, Brazil, Chile, Ireland, Germany, Luxembourg, Mexico, and New Zealand embraced both individual and State responsibility. They stated that International Law establishes the essential elements of State and individual responsibility: "These obligations entail that States and individuals are responsible and accountable for applying the law and are the ones that must be held accountable for violations" (Austria and others, 2020).

Assuring human accountability as far as possible is crucial. Nonetheless, one cannot disregard the importance of State responsibility, which is heightened in the context of AWS, as it might happen that responsibility cannot be attributed to humans.

---

individual responsibility for the employment of means or methods of warfare involving the potential use of weapons systems based on emerging technologies in lethal autonomous weapons systems in accordance with their obligations under IHL" (CCW GGE, 2019a, p. 4).

The Chairperson presented "Possible questions for the GGE to explore in 2019" which include: "• What is the responsibility of States or parties to a conflict, commanders, and individual combatants in decisions to use force involving autonomous weapons systems, in light of the principles of international law derived from established custom, from the principles of humanity and the dictates of public conscience (Martens Clause)?

• How is responsibility ensured for the use of force with existing weapons that employ or can be employed with autonomy in their critical functions? Relevant existing weapons could include types of: • Air defence weapon systems with autonomous modes or functions; • Missiles with autonomous modes or functions; • Active protection weapon systems with autonomous modes or functions; • Loitering weapons with autonomous modes or functions; • Naval or land mines with autonomous modes or functions; • "Sentry" weapons with autonomous modes or functions." (CCW GGE, 2019b, pp. 11–12).

22 The chair paper on Commonalities in National Commentaries on Guiding Principles stated that "18. Several commentaries underscored that guiding principle (i)10 reaffirmed that weapons can only ever be tools lacking agency and legal personality. It was further noted that machines are not moral agents. As such, policy measures must always address humans" (Chair of the 2020 GGE LAWS, 2020).

23 The United States Delegation commented on the Guiding Principles that "This guiding principle reflects the fundamental importance of human responsibility in using machines. The GGE should elaborate on guiding principle (b) by addressing how well established international legal principles of State and individual responsibility apply to States and persons who use weapon systems with autonomous functions. Such work could inform practical measures to promote accountability for such decisions, addressed under guiding principle (d)" (US Delegation, 2020).

2.2.2 The 2022 Shift in the GGE as Regards State Responsibility

The issue of State responsibility remained on the sidelines until the second meeting of 2022, when it started to gain momentum during the GGE, and was part of the oral debate and written submissions (Australia and others, 2022; Argentina and others, 2022a; and the Bolivarian Republic of Venezuela, 2022), the UNIDIR resource paper (Anand, 2022), and the draft report of the Chair of the 2022 Group of Governmental Experts (Chair of the 2022 GGW LAWS, 2022).[24] The following statement on State responsibility was adopted by consensus in the 2022 final report:

> For the purposes of its work, the Group recognized that every internationally wrongful act of a State, including those potentially involving weapons systems based on emerging technologies in the area of LAWS entails international responsibility of that State, in accordance with international law. In addition, States must comply with international humanitarian law. Humans responsible for the planning and conducting of attacks must comply with international humanitarian law (GGE CCW, 2022b, p. 4).[25]

The paragraph is a vital hallmark to ensure State responsibility in the context of breaches involving AWS. However, it is still only the first step, as it excluded, for instance, a statement of attribution contained in the draft report of the Chair, the UNIDIR resource paper (Anand, 2022), and the document "Principles and Good Practices on Emerging Technologies in the Area of Lethal Autonomous Weapons Systems – Proposed by Australia, Canada, Japan, the Republic of Korea, the United Kingdom and the United States" (Australia and others, 2022). Attribution will be dealt with in-depth in Part II, Chapter 4. The doors are open, and now a broad discussion on State responsibility is necessary, as will be demonstrated in Parts II and III.

We acknowledge that the CCW GGE is a diplomatic forum with limited room for legal analysis, as many experts are not international lawyers. The *locus* for its development is legal doctrine and scholarship that can further impact diplomatic debates.[26] In this book, we provide reflections on State responsibility as regards AWS that might, in the future, impact diplomatic debates.

Part I has highlighted the AWS concept, its inherent unpredictability, the accountability gap, the lack of focus on State responsibility, and the 2022 pivot

---

[24] We also acknowledge that State responsibility was part of the Chairperson's summary in the CCW GGE meeting of 2021, but was not part of the final report. See Annex III CCW/GGE.1/2021/3.

[25] For deeper comprehension, see Appendix K.

[26] This is a citation from the author's Ph.D. dissertation. During an interview for the author's Ph.D., Diego Mauri, in his personal capacity, stated that:
"In this negotiations at CCW GGE (…) international lawyers are a minority and what you can read in State's intervention or in final memorandums, is you have few and recurring legal arguments (…) there is no room for appropriate legal analysis and that forum, which understandably is more a topic for legal doctrine or legal scholarship" (Mauri, personal communication, July 5, 2022).

of the GGE regarding State responsibility. This background is necessary to address the proposed question. Part II focuses on the first clause of the question: the analysis of the ARSIWA and whether it suffices or not to ensure State responsibility for AWS violations.

Part II
International Responsibility of States and Autonomous Weapons Systems: Articles on the Responsibility of States For Internationally Wrongful Acts

While Part I has set the necessary foundations for the discussions of this book, the purpose of the chapters of this second part is to dive into the first clause of the research question:

Is the existing international legal regime on the international responsibility of States capable of addressing breaches concerning AWS? If not, what are the main gaps?

Part II will analyze possible challenges Autonomous Weapons Systems (AWS) may pose to the international responsibility of States, as stated in the Articles on the Responsibility of States for Internationally Wrongful Acts (ARSIWA). Part II will also consider whether AWS' features might challenge State responsibility due to the content of IHL and IHRL primary rules. For this purpose, we studied ARSIWA, its commentaries, archives of the drafting procedure, the current debate on the 6th committee of the UN to transform ARSIWA into a treaty (UNGA, 2019), CCW GGE documents, and the doctrine, among other sources.

Part II aims to envision possible challenges the current regime might present from the perspective of an attorney of a plaintiff, defendant, and victim State of an internationally wrongful act in the context of Autonomous Weapons Systems (AWS).

Part II Objective and Structure

Since Part II's objective is to analyze issues around AWS regarding the international responsibility of States based on ARSIWA, it follows, as far as possible, the structure of the ARSIWA. Chapter 3 deals with ARSIWA and AWS in general. Chapter 4 addresses the general principles of ARSIWA, attribution of conduct to a State, and breach of an international obligation. Chapter 5 discusses the responsibility of one State in connection with another State's action. Chapter 6 embraces circumstances precluding wrongfulness and the guarantees of non-repetition. Finally, Chapter 7 discusses specific primary issues directly related to ARSIWA: human-machine interaction, responsibility for not using AWS, weapons review, and due diligence. We will only analyze the issues in which AWS pose additional challenges. Therefore, we will not analyze the parts of ARSIWA that do not meet this condition, such as Parts III and IV of ARSIWA on "The implementation of international responsibility of States" and "General Provisions" respectively.

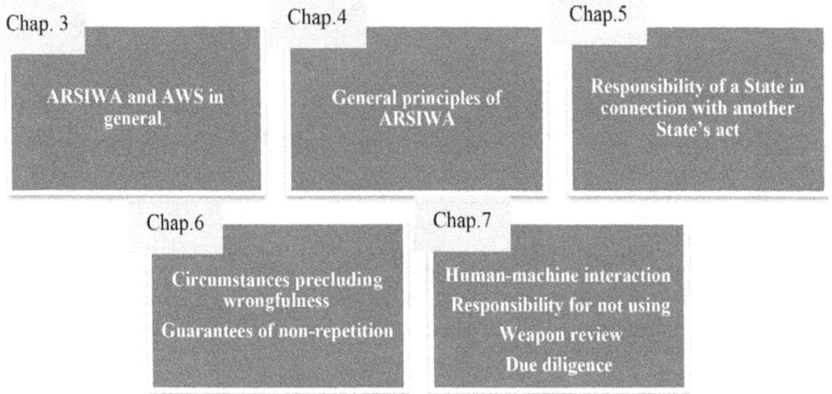

CHAPTER

3

The Articles on the Responsibility of States for Internationally Wrongful Acts and Autonomous Weapon Systems

Attempts at codification of State responsibility date back to the beginning of the 20th century. However, with the creation of the International Law Commission (ILC) in 1948, a long process began that culminated in the Articles on the Responsibility of States for Internationally Wrongful Acts (ARSIWA), adopted by the ILC in 2001 and annexed to UNGA Resolution 56/83 (Saliba and Barbosa, 2023). The ILC's final decision was to suggest that the UN General Assembly take note of the articles and not adopt them as a treaty.[1]

Despite the non-binding character of the form in which the ARSIWA were adopted, they largely reflect international custom. The articles were based on many cases from the ICJ and its predecessor, the Permanent Court of International Justice, as observed in the ILC's commentaries. The ARSIWA have heavily influenced international responsibility scholarship, teaching, learning, and thought (Bordin, 2021). International courts, tribunals, and other bodies have used the ARSIWA as the basis for decisions and heavily cited them, including controversial

1 Teles affirms that "Three factors weighted in the final decision of the Commission. First, the International Court of Justice's reliance on the draft articles before their conclusion in Gabčíkovo-Nagymaros (e.g., paras 47, 50–53) and Cumaraswamy (para 62), strengthened the argument that the articles would remain relevant even if they were not turned into a treaty. Second, the precedent of UNGA Resolution 55/153 (2001), which, upon the Commission's recommendation (here at 20, para 44) took note of the articles on Nationality of Natural Persons in Relation to the Succession of States and decided to consider them again at a later session. Third, the absence of dispute settlement provisions in the draft meant that the Commission had more flexibility in selecting the form. The dispute settlement provisions adopted on first reading in 1996 had proved controversial, and the Commission had decided to remove them from the draft. The Commission left it to the UNGA to decide whether to include a dispute settlement mechanism should a convention on the topic be adopted, while drawing the UNGA's attention to the provisions already drafted" (Teles, 2021).

clauses (Bordin, 2021; UN Secretary-General, 2019; Tams and Paddeu, 2021).[2] In sum, despite not being a binding responsibility instrument, the ARSIWA is the lynch pin of State responsibility.

> "The ILC (…) has encoded the way we think about State responsibility. It has done so in the two senses of the word 'code': by offering a systematization of the rules in this field, the ILC has left international lawyers with a code in a legal sense. But perhaps as importantly, the ILC has also provided international law with a way of thinking about responsibility, with concepts and a terminology – with a code in a lexical sense" (Tams and Paddeu, 2021).

The ARSIWA focuses on secondary rules, which are the International Law requirements for a State to be deemed responsible for violations and the consequences of that breach (Graefrath, 1984, p. 21).[3]

Considering that the ARSIWA are the building blocks of the international responsibility of States, this book treats them as the necessary framework to analyze AWS and the international responsibility of States.

The ARSIWA took over forty-five years to be concluded and were drafted without considering the new paradigm of action AI poses. AWS represent a shift in accepting decision-making by devices (Pourcel, 2018, p. 77)[4] and challenges international accountability which traditionally has been grounded on the human role model. With the advent of AI and autonomy, the human's role diminishes, and the level of control and foreseeability drops or even vanishes (Geiß, 2016).[5]

There are competing views regarding the adequacy of the existing international law norms in dealing with AWS. Thus, examining the existing regime

---

2 Tams and Paddeu highlighted that "As the high road of treaty-making and GA declarations is blocked, they have traveled by the low road: NGOs and domestic courts cite them, as do scholars, governments and the ICJ. Sometimes, notably in the early 2000s, such citations may have been in the form of simple sign posting. But over time, the Articles have sunk in; they now are the obvious reference point for any debate about State responsibility, trigger serious engagement and sometimes directly shape outcomes" (Tams and Paddeu, 2021).

3 Graefrath affirms that: "Responsibility cannot be separated from the international obligation, but the relationship of responsibility, the legal relationship produced by the violation of an international obligation between the violator and the State suffering from violation of its rights, differs distinctly from the violated legal relationship, the primary obligation" (Graefrath, 1984, p. 21).

4 Pourcel highlights this paradigm shift "En effet, da dotation en IA implique par logique un changement de paradigm et lácceptation en cohérence dúne autonomie dáction de plus em plus forte des équipements dans leur mode de functionement et, par consequente, la capacite à prenderedes decisions dans um cadre normé" (Pourcel, 2018, p. 77).

5 Geiß explains that "Traditional accountability models are typically premised on some form of control and/or foreseeability. Higher levels of autonomy in weapons systems, however, mean lower levels of control and foreseeability. Accordingly, the more autonomous a (weapons) system is, the more difficult will it be to establish accountability on the basis of traditional accountability models" (Geiß, 2016).

of responsibility is a relevant framework to test whether new rules are necessary. If the current regime implies gaps in the responsibility of States, which is one of the bases of the edifice of international law, the necessity of complementary rules becomes evident. So far, despite the ongoing debates at the CCW and the existing IHL and IHRL framework, AWS have no specific primary rules (GGE CCW, 2019a, annex IV).

Part II discusses whether the ARSIWA are sufficiently flexible to address the challenges posed by AWS. As previously stated in Part I, considering the hardships or gaps in personal accountability regarding AWS, it is even more relevant to analyze the venue of State responsibility that has remained on the sidelines so far. First, there are some introductory remarks on AWS and State responsibility. Next, specific chapters analyze challenges to State responsibility.

3.1 AWS and State Responsibility

Most scholars and NGOs regard international responsibility as unproblematic (Amoroso, 2020, p. 146).[6] They affirm that the regime of international responsibility applies to AWS. However, they miss an analysis of possible roadblocks.

Even the NGO, Human Rights Watch, which advocates for an AWS ban, in the report "Mind The Gap The Lack of Accountability for Killer Robots" (Docherty, 2015, p. 13), claimed that State responsibility could be assigned

---

6 Among the scholars that envision State responsibility as unproblematic are:

(a) Ford states that "Where there are no applicable circumstances precluding the wrongfulness of the act of the State, a breach of an international obligation generates obligations for the breaching State and rights for the injured State. The obligations on the breaching State include the obligation to "cease the wrongful conduct [and] make full reparation for the injury caused by the internationally wrongful act." Additional consequences follow if the internationally wrongful act "constitutes a serious breach by the State of an obligation arising under a peremptory norm of general international law." Thus, where a State violates an international obligation through the operation of an autonomous weapon and there is no circumstance precluding the wrongfulness of the act, then there exist several fora in which the State may be held liable. The Draft Rules also incorporate a robust regime for countermeasures which can be taken by the injured State. Judicial fora include the International Court of Justice and domestic courts; though accountability through both systems would be hindered by issues of jurisdiction." Ford acknowledges some jurisdictional challenges "Judicial fora include the International Court of Justice and domestic courts; though accountability through both systems would be hindered by issues of jurisdiction" (Ford, 2017 pp. 475–477).

(b) Schmitt, who affirms that "States can be held accountable under the laws of State responsibility should their armed forces use AWS in an unlawful manner" (Schmitt, 2015).

(c) The communication from the Commission to the European Parliament, the European Council, the Council, the European Economic and Social Committee and the Committee of the Regionsstated that "The application of AI in weapons systems has the potential to fundamentally change armed conflicts and therefore raises serious concerns and questions. The Union will continue to stress that international law, including International Humanitarian Law and Human Rights Law, applies fully to all weapons systems, including autonomous weapons systems, and that States remain responsible and accountable for their development and use in armed conflict" (European Commission, 2018, p. 8).

relatively easily.[7] This book's perspective is that the NGO missed one crucial argument in its advocacy: current international law is insufficient to ensure State responsibility for violations committed by AWS, as demonstrated in the chapters of Part II of this book, which questions this assumption and analyzes to what extent AWS challenge State responsibility.

In this sense, there is nascent scholarship underscoring that the ARSIWA need to adapt (Mauri, 2022, p. 232)[8] to address the technological development of warfare that has occurred since the U.N. General Assembly has taken note of the ARSIWA (Mačák, 2021). International law does not provide a clear path for AWS. AWS add various challenges regarding State responsibility associated with autonomy (Bruun, personal communication, June 21, 2022),[9] such as who shall bear the responsibility, mainly if a human did not perform the decisive action (Garcia, 2016, p. 111).[10]

Until July 2022, as in the scholarly works, the diplomatic debate was at a very initial stage. Among the few highlighting the issue was the NAM, which stated that the GGE should address the normative and operational framework regarding States' international responsibility for AWS's internationally wrongful acts (Bolivarian Republic of Venezuela on behalf of NAM, 2020).[11] The US emphasized the need for the GGE to address specifically State responsibility concerning AWS (US Delegation, 2020).[12] Notwithstanding recognizing the challenges, neither the scholarship nor the States at CCW presents an in-depth analysis of the main gaps regarding ARSIWA.

---

7 The report states that "While state responsibility for the unlawful acts of fully autonomous weapons could be assigned relatively easily to the user state, as discussed in Chapters 3 and 4, it would be difficult to ascribe personal responsibility for those acts" (Docherty, 2015, p. 13).

8 Mauri's perspective is that "state responsibility could be easily adjusted to the misdoings of AWS, either via a principled interpretation of existing law or – even more appropriately – through the adoption of ad hoc legal instruments establishing a regime of absolute liability" (Mauri, 2022, p. 232).

9 This is a citation from the author's Ph.D. dissertation. During an interview for the author's Ph.D., Laura Bruun, in her personal capacity, stated: "On top of the existing challenges (on State responsibility), there are some new challenges associated with autonomy" (Online, 21 June 2022).

10 Garcia reflects: "(…) Who is to be held accountable, especially if the primary actor is not a moral agent for command and responsibility? What does it do to our identity as humankind, and why forgo humans' unique capacity to engage in moral reasoning during war? In the end, questions of state responsibility abound. International law in this case provides some constraints, but it does not provide clear answers for future arms and technologies" (Garcia, 2016, p. 111).

11 NAM stated that among the elements to be considered in the normative and operational framework of AWS is "(b) The responsibility of States for internationally unlawful acts caused by lethal autonomous weapons systems" (Bolivarian Republic of Venezuela on behalf of NAM, 2020).

12 The US delegation commented "This guiding principle reflects the fundamental importance of human responsibility in using machines. The GGE should elaborate on guiding principle (b) by addressing how well established international legal principles of State and individual responsibility apply to States and persons who use weapon systems with autonomous functions. Such work could inform practical measures to promote accountability for such decisions, addressed under guiding principle (d)" (US Delegation, 2020).

3.2 The 2022 Shift in the Debate at the CCW GGE on LAWS

There was a turnaround during the CCW GGE 2022 second session, and suddenly, State responsibility gained attention (Anand, 2022, p. 17).[13] As mentioned in Part I, an initial statement on State responsibility was agreed upon in the GGE on LAWS 2022 final report:

> "(...) the Group recognized that every internationally wrongful act of a State, including those potentially involving weapons systems based on emerging technologies in the area of LAWS entails international responsibility of that State, in accordance with international law" (GGE CCW, 2022b).

Despite the importance of the statement, it only addressed the tip of the iceberg since issues such as attribution, contained in the draft proposal, were not included in the final report.

---

[13] State responsibility was raised, for example, by Cuba, France, and Argentina during the 2022 second session of the GGE on LAWS during oral debates (GGE LAWS, 2022a). Written contributions also embraced State responsibility (Australia and others, 2022; Argentina and others, 2022a; the Bolivarian Republic of Venezuela on behalf of NAM, 2022). State responsibility was expressed in Chair of the 2022 GGE, 2022 (par. 20).

CHAPTER

4

The Internationally Wrongful Act of a State: General Principles, Attribution, and Breach of an International Obligation

According to the ARSIWA (ILC, 2021c, art. 2),[1] the elements of an internationally wrongful act of a State are twofold: an action or omission is attributable to the State according to international law, and that action or omission constitutes a breach of an international obligation of the State. In this chapter, we will address both elements concerning violations committed in the context of AWS, first attribution (4.1) and then the breach of an international obligation (4.2). Next, we will discuss the damage as an unnecessary element (4.3), and AWS and the challenges of stating the time of an internationally wrongful act (4.4).

4.1 Attribution of Conduct to a State: Do AWS Call for Special Rules?

This section will analyze the attribution of an action or omission in the context of AWS to the State. Attribution is the attachment of an individual's legal situation to the State so that under international law, the action is considered an act of the State (Kolb, 2017; Condorelli and Kress, 2010).[2] We will focus

1 Note the 2001 Draft Articles on Responsibility of States for Internationally Wrongful Acts wording: "Article 2. Elements of an internationally wrongful act of a State There is an internationally wrongful act of a State when conduct consisting of an action or omission: (a) is attributable to the State under international law; and (b) constitutes a breach of an international obligation of the State." (ILC, 2001c, art. 2) is very close to the 1973 Report of the ILC: "Article 3. Elements of an internationally wrongful act of a State There is an internationally wrongful act of a State when: (a) Conduct consisting of an action or omission is attributable to the State under international law; and (b) That conduct constitutes a breach of an international obligation of the State" (ILC, 1973, art. 3).

2 According to Kolb: "The notion of attribution thus concerns this 'attachment' of a legal position from one individual to another subject: the legal fact that the action of one is considered to be the action of another. Hence, attribution can be defined as a legal operation through which acts

(*Contd.*)

on the difficult but most frequent case when dealing with AWS, of an act, at least to some extent, outside the sphere of foreseeability of the human being in charge (Scharre, 2018, p. 186; Heyns, 2014), and not on situations in which the conduct stems from a clear human intent. As previously discussed in Part I, AWS present a degree of unpredictability. The inquiry will start with the ARSIWA, its commentaries, and related documents and proceed to doctrinal writings. We will first analyze general works on attribution that do not specifically discuss AI or AWS. Next, are the writings of authors that consider AWS and attribution: first, those who deny attribution difficulties, and following, those who acknowledge the existence of attribution obstacles. States' statements, especially in the 2022 CCW GGE, will then be presented. There is no international case law on AI and attribution of international responsibility, but we will refer to relevant cases on related issues.[3]

---

or omissions of individuals performed on behalf of the State are considered legally to be acts and omissions of the latter, in view of a certain legal function (here the one of responsibility)" (Kolb, 2017, pp. 70–71).

Condorelli and Kress state that "By all accounts, none of the subjects of international law belonging to the genus of collective entities (States, international organizations, etc) is able to carry out its activities, whatever they may be, other than through individuals. 'Attribution' (or 'imputation') is the term used to denote the legal operation having as its function to establish whether given conduct of a physical person, whether consisting of a positive action or an omission, is to be characterized, from the point of view of international law, as an 'act of the State' (or the act of any other entity possessing international legal personality)" (Condorelli and Kress, 2010).

3 Regarding domestic cases, there is no case under the research terms "Artificial Intelligence" or "Autonomous Weapon(s) System(s)" o in the Oxford International Law and Domestic Courts Database (Search in 10/11/2021). (Oxford Public International Law. n.d.). There is no case under the research terms "Artificial Intelligence" "Autonomous Weapon(s) System(s)" in the Israeli Supreme Court Website (Search in 01/13/2022) (Israel. n.d.). There were two cases on the French Cour de cassation's website under the research term "intelligence artificielle" but none of them referred to responsibility as regards artificial intelligence's misdoings. One was about "la société Intelligence Artificielle" and the other "chef du service intelligence artificielle". There were no judgements under the term "systèmes d'armes autonomes" (Search in 01/13/2022) (France, n.d.). The People's Court of the People's Republic of China Supreme Court website in English has no document regarding "Autonomous Weapon(s) System(s)" and some posts under the term "artificial intelligence". The post that is relevant to the present book affirms that "A court in Shenzhen, Guangdong province, ruled recently that a work generated by artificial intelligence qualified for copyright protection." We note that the English website does not allow for decision search (Search in 01/13/2022) (People's Republic of China, 2020). In the Supreme Court of Korea, there is no case under the research term "Autonomous Weapon(s) System(s)" and three results under the research term "Artificial Intelligence," none of them relevant to the present book (Search in 01/13/2022) (Korea, n.d). We found four documents under the research "Artificial Intelligence" using the US Supreme Court advanced search for opinions and court orders. However, none of them were judicial decisions (https://www.supremecourt.gov/search_center.aspx Accessed 13 January 2022). We found no judgments under "Autonomous Weapon(s) System(s)" (Search in 01/13/2022) (US, n.d.). We found no case under the research terms "Artificial Intelligence" or "Autonomous Weapon(s) System(s)" using the UK's Supreme Court search engine (Search on01/13/2022) (UK, n.d.).

Regarding Artificial Intelligence and accountability, the Business and Corporate Litigation Committee presented the following cases:

(*Contd.*)

4.1.1 The ARSIWA and Attribution of the Conduct of AWS to the State

The ILC's commentaries on draft article 2 state that some have described attribution as a subjective element and the breach as an objective element and further argue that the ARSIWA avoid this terminology (UN ILC, 2001b, p. 34). Nonetheless, only violations that can also be subjective depending on the primary obligation are analyzed, e.g., the torture convention (Convention Against Torture and Other Cruel, Inhuman or Degrading Treatment or Punishment, 1984), and no comments are made regarding the possibility of attribution being objective.

According to the ILC′s commentaries, an act of the State must involve action or omission by a human being (UN ILC, 2001b, p. 35).[4] The commission cites the PCIJ's 1923 Advisory Opinion regarding German settlers in Poland, in which the Court stated that "States can act only by and through their agents and representatives" (UN ILC, 2001b, p. 35; UN ILC, 1973, p. 181a).[5] In a footnote,

"[Fed] Algorithmic Accountability Act (Apr 2019). Bills S 1108, HR 2231 (Apr. 2019) intended to require "companies to regularly evaluate their tools for accuracy, fairness, bias, and discrimination."

[NJ] New Jersey Algorithmic Accountability Act (May 2019). Require that certain businesses conduct automated decision systems and data protection impact assessments of their automated decision system and information systems.

[CA] AI Reporting (Feb 2019). Require California business entities with more than 50 employees and associated contractors and vendors to each maintain a written record of the data used relating to any use of artificial intelligence for the delivery of the product or service to the public entity.

[WA] Guidelines for Gov't Procurement and Use of Auto Decision Systems (Jan 2019). Establish guidelines for government procurement and use of automated decision systems in order to protect consumers, improve transparency, and create more market predictability.

[NY] NYC (Jan 2018). —"A local law in relation to automated decision systems used by agencies" (Int. No. 1696–2017) required the creation of a task force for providing recommendations on how information on agency automated decision systems may be shared with the public and how agencies may address situations where people are harmed by such agency automated decision systems" (Business and Corporate Litigation Committee, 2021).

There are, for instance, domestic cases on Tesla′s autopilot, nonetheless, the company argues that those systems are not autonomous and should be supervised by the driver (Murdock, 2020).

In September 2021, the UK Court of Appeal ruled that AI cannot be listed as an inventor on a patent application (UK Court of Appeal, 2021).

4 ARSIWA commentaries explain that "(5) For particular conduct to be characterized as an internationally wrongful act, it must first be attributable to the State. The State is a real organized entity, a legal person with full authority to act under international law. But to recognize this is not to deny the elementary fact that the State cannot act of itself. An "act of the State" must involve some action or omission by a human being or group: "States can act only by and through their agents and representatives." The question is which persons should be considered as acting on behalf of the State, i.e. what constitutes an "act of the State" for the purposes of State responsibility" (UN ILC, 2001b, p. 35). The commentaries cite German Settlers in Poland, Advisory Opinion, 1923, P.C.I.J., Series B, No. 6, p. 22.

5 See also ILC's 1973 explanation: "In the last analysis, therefore, conduct regarded as an "act of the State" can only be some physical action or omission by a human being or group of human beings. Hence the necessity of establishing when and how an "act of the State" can be discerned in a given action or omission. In other words, it is a question of determining by whom and in what circumstances these actions or omissions must have been performed for them to be attributable to the State" (UN ILC, 1973, p. 181).

the ILC commentaries refer to paragraph (1) of the commentary to article 3 of The Report of the International Law Commission (ILC) of 1973. The 1973 report states that attribution, which is generally called a subjective element, consists of "(...)conduct that must be capable of being attributed not to the human being or group of human beings which actually engaged in it, but to the State as a subject of international law"(UN ILC, 1973). Notably, human beings are the ones who engage in the conduct regardless of whether such conduct is attributable to them or not. This framework demonstrates a long-lasting understanding of the need for human action for an act or omission to be attributed to the State. Therefore, under the ARSIWA, and the current state of international law, the general rule is that attribution is subjective, requiring a human agent´s conduct unless specific primary rules foresee differently.

Article 4 of the ARSIWA states that the conduct of State organs is considered an act of that State under international law. Organ embraces persons and entities that are so regarded by the State's internal law (UN ILC, 2021c, art. 4). Article 5 addresses power delegation, so that the conduct of persons or entities that are not organs of the State but that, according to the law of that State, are empowered to exercise elements of governmental authority shall be considered an act of that State (UN ILC, 2021c, art. 5). The commentaries again lean towards a human action paradigm since both natural and legal persons are mentioned (UN ILC, 2021b)[6] and, as decided by the ICJ in the aforementioned German Settlers in Poland Advisory Opinion, States need their agents and representatives to act.

AWS actions can result from a human being's action or omission. Still, as explained in Part I, due to the discretion and inherent unpredictability of AWS by reason of algorithmic complexity or the interaction with the environment, they might behave in venues unpredictable to the human beings who programmed or deployed them. In such hypotheses, within the current framework of the ARSIWA, it is hard to find a direct human action to attribute responsibility to the State. The causal chain is likely to be broken. The course of action can be attributed to a machine whose functioning cannot be attributed to any human person. Under current international law, AWS are not considered a State organ since organs are composed of humans and not solely machines. In short, considering AWS´ unpredictability, the internationally wrongful act might not be attributable to any person or entity.[7]

---

6 ILC´s commentaries state that "(12) The term "person or entity" is used in article 4, paragraph 2, as well as in articles 5 and 7. It is used in a broad sense to include any natural or legal person, including an individual office holder, a department, commission, or other body exercising public authority, etc." (UN ILC, 2021b).

7 Note that AWS unpredictability and the following responsibility cannot be compared to the unpredictability of landmines. First, in relation to landmines, there is a series of obligations arising from treaties that provide for the obligation to remove mines or install warnings. In this sense (Protocol II, on the Use of Mines, Booby-Traps and Other Devices). (1998) and the (Anti-Personnel Landmines Convention, 1997). Furthermore, landmines represent a "physical unpredictability" whereas AWS are unpredictable both due to interaction with the environment and due to AI and the complexity of algorithms.

In conclusion, the paradigm of action in the ARSIWA is human-centered, which means that attribution of international responsibility has so far been the attachment of acts or omissions of a human being(s) to the State.

State responsibility is also a rule of customary IHL (ICRC, n.d.e, Rule 149). IHL is based on a human action paradigm, and Article 3 of the 1907 Hague Convention and Article 91 of Additional Protocol I, foresee State responsibility for “all acts committed by persons forming part of its armed forces” (Hague Convention IV, 1907, art; 3; Additional Protocol I, 1977, art. 91).

> Existing international law, including international humanitarian law, while still applicable, is insufficient because its fundamental rules regarding the use of force were designed when humans made value judgments notably vis-à-vis the principles of distinction, proportionality, precautions in attack and military necessity at the moment of the application of force (Chile and Mexico Delegations, 2022).

As can be observed, State responsibility arises from human actions. International law, including the AIRSIWA and IHL, so far do not include autonomous devices, such as AWS or autonomous vehicles, as members of its armed forces.

4.1.2 Doctrine on the ARSIWA and the Human Action Paradigm

Doctrinal works on State responsibility also corroborate the human action paradigm of the ARSIWA.

4.1.2.1 General Doctrine on the ARSIWA and Attribution

Condorelli and Kress state that only through individuals do States and International Organizations manage to carry out their activities and do not discuss actions by AI (Condorelli and Kress, 2010). Sassòli and Nagler also emphasize that only humans can act and that human actions are attributable to the State (Sassòli and Nagler, 2019, pp. 173–174).[8] Similarly, Kolb presents a human-centered view without considering AI and argues that States “can act only though persons” (Kolb, 2017, p. 70)[9] and addresses individuals’ actions or omissions.

8 Sassòli and Nagler emphasize that “Even though only human beings can act in the real world, many rules of international law, including those on whether an armed conflict is international, refer to States. Therefore, one must attribute the actions of those who acted to a State in order to determine whether one State has used force against another State” (Sassòli and Nagler, 2019, pp. 173–174).

9 Kolb affirms that “The State, being a moral person without any physical existence, can act only through persons whose conduct is performed on behalf of the State. Such persons are called organs or agents, according to their precise link to the State. The action or omission of such individuals as are acting on behalf of the State is then considered, from the legal point of view, to be an action of the State. There is no representation or mandate. The action is not considered to be an action of the individual later transferred to the State, but merely and exclusively an action of the State” (Kolb, 2017, p. 70).

This short overview shows that the general doctrinal works on the attribution of international responsibility of States do not consider the new paradigm of action posed by AI, meaning AI developing tasks formerly developed only by humans, and only consider the possible attribution of human action to States.

4.1.2.2 Doctrine on the AWS and Attribution

We will turn to the few scholarly works that touch upon AWS and attribution. Scholarship can be, in broad terms, divided into two mental tracks: The first deems that there is no attribution problem and that, in broad terms, the act was done by State organ fielding, which we claim does not find support in current international law´s human action paradigm. The second, which resonates with ARSIWA and international law in general, is that there are attribution challenges under the current international law framework. It exposes, even if in nascent terms, that it is necessary to trace the conduct back to the State through the actors who caused the breach, which means finding a human action to attribute the conduct to the State.

Regarding the first mental track, some scholars state that AWS do not create additional attribution challenges. We disagree with the following viewpoints of authors such as Stürchler and Siegrist, Geiß, Hammond, and Ford.

Stürchler and Siegrist claim that States are responsible for AWS misdoings, even if they are unexpected or caused by a malfunction, as attribution rules apply to AWS just like any other weapon. Their premise is that AWS lack legal personality and thus cannot be deemed an agent. Accordingly, attribution is not grounded on the nature or features of AWS but on the legal status of the person deciding its deployment (Stürchler and Siegrist, 2017). In a similar perspective, Geiß ground attribution on the organ that decides to employ AWS (if a State organ decides to deploy AWS, all weapon actions are attributable to the State) (Geiß, 2016).[10] Stürchler and Siegrist, and Geiß's statements do not scrutinize whether the ARSIWA require human activity or includes AI´s actions. In our view, their perspective on attribution does not resonate with the current international responsibility regime. It could be a way out of *de lege ferenda* as an evolutionary interpretation of the ARSIWA. As they stand, the ARSIWA do not embrace such an extensive interpretation of attribution and require actual human action. Moreover, the authors do not address the "many hands problem" that many persons might be involved in the AWS deployment, and not all of them might be exercising government authority.

---

[10] Geiß stated at the CCW GGE that "In accordance with general rules on state responsibility, a State is responsible for internationally wrongful acts that are attributable to it. No particular legal challenges arise with regard to the attribution of acts committed by autonomous weapons systems. For as long as human beings decide on the deployment of these systems, accountability can be determined on the basis of established rules on attribution. Thus, if a member of the armed forces (i.e., a state organ) of State A decides to deploy a robot on a combat mission, all activities carried out by the robot are attributable to that State. The mere fact that a weapons system has (some) autonomous capabilities does not alter this assessment" (Geiß, 2016).

Hammond states that since military and intelligence agencies are State organs, considering that AWS operate under their authority, their actions can be attributable to the State (Hammond, 2015, pp. 668–669). The author does not consider the human action paradigm that linchpins attribution under the ARSIWA. Hammond does not explain how to ensure AWS act under the organ authority since their main feature is autonomy, their pace of action out-speeds humans, and they are inherently unpredictable due to the complexity of algorithms and interactions with the environment.

Ford confers to AWS the same juridical treatment as non-autonomous weapons, disregarding AWS unique features. In Ford's view, wrongful acts are attributable to the State even if AWS act beyond the sphere of foresee ability and scope of the person who deploys one (Ford, 2017, pp. 430–431).[11] This attribution claim does not specify if the actions are attributable to the State who owns the weapon, the one who deploys the weapon, or the one who programmed the weapon? According to Ford, attribution is grounded in article 5 of the ARSIWA:

> "The Articles do not, of course, consider the potentiality of autonomous weapons systems. Article 5 does, however, consider a situation in which "a person or entity empowered by the law of that State to exercise elements of the governmental authority shall be considered an act of the State under international law." This article might be regarded as autonomous weapons, which are systems that have been "empowered" by the State (e.g., activated) in order to "exercise elements of the governmental authority" (e.g., conduct combat operations)" (Ford, 2017, pp. 475–477).

He further states that if Article 5 applies to AWS, the liability extends both to foreseen and unforeseen acts under Article 7 of the ARSIWA, which deals with *ultra vires* acts (Ford, 2016, p. 477).[12]

This book disagrees that Article 5 of the ARSIWA (ILC, 2001c, art. 5) is flexible in attributing AWS conduct to States. If, in Ford's view, AWS should be treated like other weapons, how can they be deemed a person or entity exercising elements of governmental authority and not as mere objects? If AWS are, in his words, "at most organ or agent of the State," why base attribution on article 5 that talks about persons or entities that are not organs of the State under article 4? As previously stated, Articles 4 and 5 of the ARSIWA are rooted in a human

[11] Ford's perspective is that "In the hands of a State, an autonomous weapon system, like a non-autonomous weapon system, is merely an instrument for the exercise of State authority. Even where the system is acting with extreme levels of autonomy, it is—at most—an organ or agent of the State whose actions are attributable to the State. Actions will be attributable even where the system is acting in an entirely unpredictable manner and beyond the scope of the initial deployment" (Ford, 2017, pp. 430–431).

[12] Ford's claim is that "If this article is read to trigger State responsibility for an autonomous weapons system activated by a State, the attendant liability will extend to acts both anticipated and unanticipated. Article 7 makes clear that State responsibility lies even where the agent of the State is acting ultra vires. The commentary makes clear that a State will be responsible "even if the organ or entity acted in excess of authority or contrary to instructions" (Ford, 2017, p. 477).

action paradigm of natural or legal persons (UN ILC, 2001b, p. 41, par. 12). Moreover, the ARSIWA commentaries emphasize that the aim of Article 5 is to embrace parastatal entities and former State corporations (UN ILC, 2001b, p. 42),[13] a category that does not encompass AWS. Article 5 comprises a narrow category since international law must expressly authorize the action/omission as an exercise of public authority (UN ILC, 2001b, p. 43).[14] Ford does not explain how AWS can be analogized to parastatal entities or former State corporations.

In short, it is claimed that Ford´s statements on attribution and articles 5 and 7 of the ARSIWA are unsustainable under current international law. They can only be a proposal *de lege ferenda* requiring a State's consent or an evolutionary interpretation of the ARSIWA.

Mauri touches upon attribution under ARSIWA. He questions if there would be attribution problems if AWS allowed a course of action without human intervention (Mauri, 2022, p. 215).[15] He further distinguishes "between the conduct consisting in deploying AWS and the conduct consisting in selecting and engaging an impermissible target: while the former is certainly attributable to the state, the latter may raise some doubts" (Mauri, 2022, p. 2015). Nevertheless, he then objects to this perspective by arguing that such conduct is attributable under article 4 of the ARSIWA, deeming AWS as organs of the State, and further claiming that unpredictable human actions would be "ultra vires acts" under article 7 of the ARSIWA (Mauri, 2022, p. 2015).[16] Mauri´s claim could be a possibility *de lege ferenda*, as, under the current framework of the ARSIWA, which is

13 ILC comments that "The article is intended to take account of the increasingly common phenomenon of parastatal entities, which exercise elements of governmental authority in place of State organs, as well as situations where former State corporations have been privatized but retain certain public or regulatory functions" (UN ILC, 2001b, p. 42).

14 According to ILC "For the purposes of article 5, an entity is covered even if its exercise of authority involves an independent discretion or power to act; there is no need to show that the conduct was in fact carried out under the control of the State. On the other hand, article 5 does not extend to cover, for example, situations where internal law authorizes or justifies certain conduct by way of self help or self-defence; i.e. where it confers powers upon or authorizes conduct by citizens or residents generally.
The internal law in question must specifically authorize the conduct as involving the exercise of public authority; it is not enough that it permits activity as part of the general regulation of the affairs of the community. It is accordingly a narrow category" (UN ILC, 2001b, p. 43).

15 Mauri questions, "If a particular course of action is deliberated absent any human intervention, as the operator remains somewhere in the loop, could the argument be made that, technically, the ensuing violation is not imputable to the operator and thus, pursuant to the criteria outlined by the ARSIWA, attributable to the state?" (Mauri, 2022, p. 2015).

16 Mauri´s perspective is that "from a systematic perspective it can be observed that existing criteria for attributing responsibility are capable of linking the impugned conduct with the state that has decided to deploy AWS. Autonomous weapons systems would be regarded as 'organs' of that state, exercising executive powers par excellence—namely, enforcing law and conducting hostilities. On a general level, article 4 ARSIWA remains applicable. As for the attribution of the misdoings of AWS which the human operator did not have a chance to halt or prevent, these would qualify as ultra vires acts pursuant to article 7 ARSIWA, as those 'organs' (AWS) acted, in a given scenario, 'contrary to instructions' received by the human operator before deployment" (Mauri, 2022, p. 2015).

grounded on a human action paradigm, AWS cannot be deemed as organs of States. In the next section dealing with the 2022 statements at the CCW GGE, it will be demonstrated that a similar proposal was made by the chair and some States but was not agreed upon in the final report.

Belonging to the second mental track, Rebecca Crootof acknowledges that clarifying the applicability of existing law is necessary to hold States accountable and then affirms that at a doctrinal level, AWS beaches "should be simply attributed to the State fielding" (Crootof, 2016, p. 1391). Crootof insightfully scrutinizes analogies of AWS to weapons, combatants, child soldiers, and animal combatants and states that all of them misrepresent the special legal features of AWS (Crootof, 2018, pp. 79–80).[17] Despite acknowledging that using analogy implies the risk of limiting the thinking about AWS and detaining the development of specific regulations (Crootof, 2018, p. 79), she proposes that animals would be the best analogy for AWS. They "are not quite weapons" as they can act independently and unpredictably, and they "are not quite combatants" considering that they have no *mens rea* and cannot learn the law of armed conflict. Their autonomy is moderated by training.[18]

We claim that the analogy with animals makes explicit the attribution challenges. First, as with AWS, there is no international regulation or case law on the attribution of acts of animals to the State. Second, an animal´s conduct must be linked to human action to be attributable to the State. If a State uses dogs on the battlefield, State responsibility for any damage caused by the animals

[17] According to Crootof "All of the aforementioned analogies misrepresent legally salient traits of autonomous weapon systems: the weapon analogy minimizes their capacity for independent action, while the combatant, child soldier, and animal analogies overemphasize it. As with definitions that would lump landmines and Skynet in the same category, the weapon analogy "almost certainly [misses] the essence of what is new about autonomous weapons." Autonomous weapon systems are fundamentally different from inert, automated, and semi-autonomous weapons because of their ability to independently select and engage targets. In addition to the moral questions raised by partially delegating life and death decisions to algorithms, this capacity raises legal issues that simply are not addressed by the existing law governing weapons. Meanwhile, the combatant, child soldier, and animal analogies inappropriately characterize autonomous weapon systems as fully independent entities. Unlike combatants, child soldiers, and animals—who may be coerced or tricked but nevertheless retain a fundamental degree of autonomy—autonomous weapon systems may have their ability to act independently sharply curtailed. The actions of autonomous weapon systems are based on algorithms, and algorithms may be rewritten. Autonomous weapon systems may be operated solely in semiautonomous or even automatic modes by their deployers, or they may be hacked and completely controlled by an enemy" (Crootof, 2018, pp. 79–80).

[18] Crootof explains that "In many ways, animals are the best analogy for autonomous weapon systems. They are not quite weapons, insofar as they are capable of independent and unpredictable action. They are not quite combatants: they cannot be taught the law of armed conflict, and they cannot act with the requisite *mens rea* for criminal liability. Their independence is tempered through extensive training; their propensity for unpredictable action is addressed through limited use. While the animal analogy is conceptually useful, there is no international "law of animal combatants." The lack of explicit prohibitions or regulations suggests that sovereign states may do what they wish, providing little guidance regarding appropriate international regulatory standards for autonomous weapon systems" (Crootof, 2018, p. 78).

will depend on the sphere of foresee ability of a human being. If someone stole the dog and used it against a third State, would the conduct still be linked to the person who fielded the dog and therefore attributed it to the State that owns it?

It is relevant to stress some differences between AWS and animals. Animals have an innate instinct of self-preservation, and also feel fear, characteristics not possessed by AWS. One of the advantages of AWS is that they are emotionless and have no sense of self-preservation, which contributes to their preciseness (Arkin, 2009, p. 30). Many defenders of AWS state that they are expected to outperform humans, especially in specific tasks, since they might be capable of processing huge amounts of data. Animals are not expected to perform better than humans since they have rationality far more limited than humans and cannot analyze extensive data. Animals, differently from AWS, do not present the "many hands problem" that contributes to the accountability gap. With AWS, multiple people might be involved in the chain of action, so none of them can be responsible for the violation (Amoroso, 2020, p. 123).

Considering the unique features of AWS, even if they were analogized with animal combatants, attribution of such conduct would be challenging, as the current state of international law requires human action. We agree with Crootof's warning on the risk of analogies as "they restrict our ability to think imaginatively about a new technology" (Crootof, 2018, p. 80). A counterpoint is Bruun's vision that we should not make analogies, as AWS area weapon, a tool, a capability: "It is just important to stress that AWS are not replacing humans in the use of force decisions but that they are just changing how humans make these decisions" (Bruun, personal communication, June 21, 2022).[19] The remaining question is how to deal with attribution in this context of a paradigm shift in how decisions are made.

Also belonging to the second mental track and exposing the challenges AWS pose to accountability, Laura Dickinson presents a taxonomy that analyzes six elements to identify any relation of responsibility. The first element is who is liable (Dickinson, 2018, pp. 18–19). AWS might be the harm inflictors, the causal agents, but they do not fit the concept of "who." Considering the autonomy feature, paradoxically, the "who" might be many persons—the programmer, the deployer, and the manufacturer—but at the same time, none of them. Human action is so pulverized and apportioned that, alone, none of them can be considered the "who." It is the aforementioned "many hands problem." There are many people involved, but none of them can be framed as the responsible human agent (Dickinson, 2018,

[19] This is a citation from the author's Ph.D. dissertation. During an interview for the author's Ph.D., Laura Bruun, in her personal capacity, stated that "It is very important, at least for the way I understand it, not to make any analogies to humans, or animals or also any other living organs. In my understanding it is a weapon, it is a capability. It is a means that humans can use to effectuate certain goals. But it is nothing but a tool and a capability. It is just important to stress that AWS are not replacing humans in the use of force decisions, but that they are just changing how humans make these decisions" (Bruun, personal communication, June 21, 2022).

pp. 18–19).[20] Nonetheless, Dickinson does not discuss the attribution issues under the ARSIWA. Similarly, Denise Garcia touches upon the possibility of attribution challenges in the context of full automation (Garcia, 2016).

Garcia and Hortal step further into the attribution of responsibility to States and affirm that if AWS violate international law and the wrongful act is not done by a human being, it is hard to apply the international law rules of attribution of State responsibility which are grounded on human beings' behavior (Espada and Hortal, 2013, p. 12). The authors question if a State can be responsible when one of its organs or persons whose conduct might be attributable to it sends or orders the AWS. Or if States are liable if its organs or persons whose behavior might be attributable to it designed the AWS? Or is the State responsible if its army utilizes the AWS, its operating units, or a person whose conduct might be attributable to it? (Espada and Hortal, 2013, pp. 10–11). The authors emphasize that considering the gaps in the attribution of responsibility that AWS encompass, the topic must be the object of a thoughtful study that prepares the applicable juridical framework (Espada and Hortal, 2013, p. 12).[21] Despite opening the debate and acknowledging the necessity of studying the attribution of responsibility to States for AWS misdoings, Garcia and Hortal do not further explore State responsibility regarding AWS under the ARSIWA.

In summary, the discussions of this section demonstrate that attribution under the ARSIWA are based on a human action paradigm, according to the ILC′s commentaries and drafting documents and literature on the topic. There is scarce literature on AWS and attribution, and the topic remains almost unexplored. Moreover, the specific literature that argues for no attribution challenges does so without scrutinizing the ARSIWA. Thus, under the current regime of international responsibility of States, as stated by the ARSIWA, there might exist an attribution gap of AWS-related breaches that are not directly linked to a willful human action.

4.1.2.3 The 2022 GGE Discussions on Attribution

As mentioned in Part I, before 2022, when State responsibility came under scrutiny and was proclaimed in the CCW GGE's final report (GGE CCW, 2022b),[22] the

20 Dickinson cites (Mashaw, 2006) and states that "Mashaw's rubric helps us to see that that the use of autonomous weapons systems poses a fundamental challenge to the entire conception of accountability, because the autonomous weapons systems themselves can be causal agents that inflict harm, yet they do not fit neatly within the first box in the framework: they do not clearly qualify as a "who." The "who" could of course be the various human beings involved in the operation of these systems, but because of the autonomy of the weapons it is not clear that those human beings qualify as an appropriate "who" either" (Dickinson, 2018, pp. 18–19).

21 Espada and Hortal claim that:"En todo caso, estos Sistemas de Armas deben, también por las lagunas que se plantean respecto de la atribución de responsabilidad de sus eventuales violaciones del Derecho internacional, ser objeto de un profundo estudio que prepare el cuadro jurídico que les sea aplicable" (Espada and Hortal, 2013, p. 12).

22 According to the CCW GGE ′s final report: "For the purposes of its work, the Group recognized that every internationally wrongful act of a State, including those potentially involving weapons systems based on emerging technologies in the area of LAWS entails international responsibility of that State, in accordance with international law. In addition, States must comply with

(*Contd.*)

issue was seldom raised. The statement was a significant breakthrough but is still insufficient to ensure State responsibility, as, for instance, it excluded attribution.

The draft report of the Chair proposed to include attribution in the following terms:

> "The conduct of a State's organs such as its agents and all persons forming part of its armed forces, is attributable to that State, including any such acts and omissions involving the use of a weapons system based on emerging technologies in the area of LAWS, in accordance with applicable international law" (Chair of the 2022 GGE, 2022).

The chair´s proposal followed similar statements contained in the Principles and Good Practices on Emerging Technologies in the Area of Lethal Autonomous Weapons Systems (Australia and others, 2022),[23] and The Roadmap Towards New Protocol on Autonomous Weapons Systems, which stated that acts and omissions involving the use of AWS are included within the framework of a State's organs being attributable to the State (Argentina and others, 2022a).[24] Therefore, those documents expanded the scope of article 4 of the ARSIWA, by including AWS as State organs.

Had this proposal been adopted, a step toward building customary international law as regards AWS and the ARSIWA's Article 4 would have been taken. However, as it currently stands, the ARSIWA is still based on a human action paradigm. Embracing AWS within the concept of State organs under the ARSIWA's Article 4 is a possibility *de lege ferenda*.

4.1.2.4 A Spectrum: Task Autonomy X Personal Autonomy

In the previous section, we argued that the ARSIWA is based on a human action paradigm. To better comprehend the challenges of attributing AWS misdoings to States, two lines of thoughts on autonomy, personal autonomy,

---

international humanitarian law. Humans responsible for the planning and conducting of attacks must comply with international humanitarian law" (GGE CCW, 2022b).

23 The document addressed attribution by stating that "The conduct of a State's organs such as its agents and all persons forming part of its armed forces, is attributable to the State. This includes any such acts and omissions involving the use of a weapons system based on emerging technologies in the area of LAWS, in accordance with applicable international law."
"18.State Responsibility. Under principles of State responsibility:
a. Every internationally wrongful act of a State, including such conduct involving the use of a weapons system based on emerging technologies in the area of LAWS, entails the international responsibility of that State.
b. The conduct of a State's organs such as its agents and all persons forming part of its armed forces, is attributable to the State. This includes any such acts and omissions involving the use of a weapons system based on emerging technologies in the area of LAWS, in accordance with applicable international law" (Australia and others, 2022).

24 The document: "21. Reaffirm that the conduct of a state's organs such as its agents and all persons forming part of its armed forces, is attributable to that state. In accordance with IHL, IHRL, and ICL, this includes any such acts and omissions involving the use of AWS" (Argentina and others, 2022a).

and task autonomy, already briefly presented in Part I, Chapter 1, will be discussed. It is not about mixing legal analysis with philosophical musings but about recognizing that legal norms are often related to philosophical roots. This interdisciplinary comprehension can shed light on the challenges faced in the legal field: autonomy is not a univocal concept.

On one end of the spectrum is personal autonomy, which is grounded in singularity and conveys that a person, entity, or system is autonomous if it can have free will, consciously think through, and decide the action it should take, in the Enlightenment tradition (Weber and Suchman, 2016, p. 79). On the other end of the spectrum, task autonomy is grounded on goal-oriented behavior and focuses on the capacity of a person, entity, or system to develop a task without external input regularly (Tamburrini, 2016, pp. 124–125).

As previously discussed, the ARSIWA is based on human action, and thus the paradigm behind the ARSIWA leans toward the human-based illuminist agency conception grounded on personal autonomy.[25] In this regard, the case *Prosecutor v. Germain Katanga*, mentions the State's obligation under the ARSIWA to make reparation and requires personal foresee ability (U.N. Secretary-General, 2019).[26]On the other hand, most of the conceptualizations of AWS, such as the

---

[25] From our bibliographic review, it is not possible to affirm if the ILC was interested in or discussed philosophical issues of "autonomy," but, as argued above, it is possible to affirm that the ARSIWA are based on a human action paradigm. Therefore, it is reasonable to state they gravitate toward a human action paradigm as a rule. However, we acknowledge that in some exceptional situations, personal autonomy might not be present, but State responsibility still ensues. For instance, if a police officer commits atrocious human rights violations in the exercise of their duties, even if the agent is very seriously mentally ill, responsibility would be attributed to the State. Arguably, this may affect their personal autonomy (and, to a certain extent, their individual responsibility). In short, the ARSIWA presupposes human action for attribution, leaning towards personal autonomy of the agent, as a general rule, even if State responsibility can be triggered if there is human action but not personal autonomy.

[26] The document (U.N. Secretary-General, 2019) cites (ICC, 2018, par. 17 and note 36) and states that "In Prosecutor v. Germain Katanga, the Trial Chamber cited the commentary to article 31 of the State responsibility articles when finding that "if the person who committed the initial act could not have reasonably foreseen the event in question, the initial act cannot be considered to be the proximate cause of the harm suffered by the victim and, consequently, the person who committed the initial act cannot be held liable for the harm in question" (U.N. Secretary-General, 2019).

In this regard, ILC states that "Thus, causality in fact is a necessary but not a sufficient condition for reparation. There is a further element, associated with the exclusion of injury that is too "remote" or "consequential" to be the subject of reparation. In some cases, the criterion of "directness" may be used, in others "foresee ability" or "proximity". But other factors may also be relevant: for example, whether State organs deliberately caused the harm in question, or whether the harm caused was within the ambit of the rule which was breached, having regard to the purpose of that rule. In other words, the requirement of a causal link is not necessarily the same in relation to every breach of an international obligation. In international as in national law, the question of remoteness of damage "is not a part of the law which can be satisfactorily solved by search for a single verbal formula". The notion of a sufficient causal link which is not too remote is embodied in the general requirement in article 31 that the injury should be in consequence of the wrongful act, but without the addition of any particular qualifying phrase" (UN ILC, 2001b, pp. 92–93 par 11).

US DOD's, Stop Killer Robots, ICRC, and the provided definition, lean toward task autonomy, which means that a system is autonomous if it can regularly develop a task without external intervention or input. Thus, personal autonomy will hardly be helpful concerning AWS, as only conscious individuals can be personally autonomous (Tambirrini, 2016, pp. 124–125).[27]

New technologies challenge the Enlightenment premises, such as individual agency, cognition, and action (Weber and Suchman, 2016, p. 76).[28] This divergence of paradigms of autonomy of the ARSIWA and AWS seems to be one of the causes of attribution problems. While the ARSIWA require human action and judgment, AWS focus on task accomplishment (Weber and Suchman, 2016, p. 76).[29]

Considering that the ARSIWA lean toward personal autonomy based on human action, to embrace task autonomy, specific primary rules regarding AI in general, or autonomous weapons in specific, must be developed. One could also advocate for an interpretative evolution of the ARSIWA to embrace task autonomy. Nonetheless, it requires State agreement or that customary international law embraces this new interpretative solution. Another option is to resort to an interpretive extension, by analogy, of the scope of Article 4. However, as will be shown in 8.2.1.1. "Analogical interpretation possible framework in the context of AWS," analogy would be too far a stretch for current international law.

The States' proposals presented in section 4.1.2.3 to include AWS misdoings within the ambit of State organs, under article 4 of AIRSIWA, sparks this paradigm shift toward the task autonomy side of the spectrum but still needs to be agreed upon unanimously by States at the GGE or another forum.

27 Tamburrini states that "Philosophical definition of what one may aptly call personal autonomy (or p-autonomy) are hardly useful in this circumstances, insofar as only conscious individuals, who are additionally assumed to be free and capable of acting in their genuine intentions, are p-autonomous. (...) More pertinent and informative for the purpose of defining and identifying AWS is the idea of task autonomy (t-autonomy), which is construed here as a three-place relationship between a system S, a task t, and another system S". Roughly speaking, a system S is said to be autonomous at some task t from another system S′ (S′ is t-autonomous from S′) if S accomplishes it regularly without any external assistance or intervention by S′" (Tamburrini, 2016, p. 124 e 125).

28 Suchman and Weber, state that "(...) while cybernetics and behavior-based robotics challenge the premises of individual agency, cognition, communication and action that comprise the Enlightenment tradition, they also reiterate aspects of that tradition in the design of putatively intelligent, autonomous machines" (Weber and Suchman, 2016, p. 76).

29 Lucy Suchman and Jutta Weber argue for a paradigm shift in the concepts of agency and autonomy from attributes that are inherent to the acting entities to a focus on material practices: "This work suggests a shift in conceptions of agency and autonomy, from attributes inherent in entities to effects of discourse and material practices that either conjoin humans and machines or delineate differences between them. This shift leads in turn to a reconceptualization of autonomy and responsibility, as always enacted within rather than being separable from, particular human-machine configurations" (Weber and Suchman, 2016, p. 76). This rearrangement slides the concepts of autonomy and responsibility that are no longer necessarily separable from human-machine interactions. Nonetheless, this shift and a task autonomy paradigm of action have not been discussed by the international community during the process of drafting the ARSIWA. Moreover, such a slide has not so far been incorporated by the ARSIWA nor discussed by the 6th committee that deals with turning the ARSIWA into a treaty.

4.2 The Breach of an International Obligation

The breach of an international obligation of the State is the second element of the internationally wrongful act (ILC, 2001c, art. 2). It is an act of the State in non conformity with international obligation requirements, regardless of the origin or character of the obligation (ILC, 2001c, art. 12). The ARSIWA does not require intent, and "In the absence of any specific requirement of a mental element in terms of the primary obligation, it is only the act of a State that matters, independently of any intention" (UN ILC, 2001b, p. 36).

Nonetheless, it is essential to discuss if, due to many of the IHRL and IHL primary rules, it is necessary that AWS-related actions are willful or negligent for State accountability to ensue. In cases where the answer is positive, State responsibility faces the same challenges as individual criminal liability in the context of AWS.

> "Inaccurate, specifically, is the idea whereby State responsibility for IHL violations (or, where applicable, for arbitrary deprivations of life under IHRL) would not require the ascertainment of a mental element. While it is true that "intent" or "fault" do not count as constitutive elements of internationally wrongful acts under "secondary" customary norms on State responsibility, it is also generally acknowledged – including by the ILC itself – that the relevance of a culpable mental element may be envisioned by the "primary" norm whose alleged violation is at stake" (Amoroso, 2020, p. 147).

Many of the existing IHL and IHRL primary obligations that may be applicable in breaches concerning AWS require *mens rea,* which means a willful state of mind (Geiß, 2015, p. 23).Therefore, AWS's unpredictable misconduct may hinder the objective element (breach of an international obligation) and, thus, be a roadblock to attributing State responsibility. In this context, the hardships in ascribing individual responsibility based on a culpable state of mind are also a problem for ensuring State responsibility (Amoroso, 2020, p. 146).

Concrete examples of primary norms requiring intent are the IHL principles of distinction, proportionality, and the IHRL prohibition on arbitrary deprivation of life. The humanitarian law principle of distinction, which prohibits direct attacks against the civilian population, for instance, requires a mental state. If an attack causes unintended civilian casualties due to unexpected technical malfunctions, it cannot be regarded as a violation of the principle of distinction since it was not directed against civilians (Amoroso, 2020, p. 147). Similarly, a breach of the principle of proportionality requires *mens rea.* Under IHRL, a breach of the prohibition on arbitrary deprivation of life also requires intent. In fact, "Whether civilian casualties are merely tragic accidents or war crimes depends on *mens rea*, and the intentions of those who order attacks and carry them out" (Asaro, 2014).

In sum, a vacuum might emerge with AWS, meaning no breach of international obligations, in situations where primary norms require a mental element. To address this issue, we question if *de lege lata*, intentionality could be deemed as recklessness or negligence for State responsibility in the context of AWS, or *de lege ferenda* presumed through strict liability (as will be discussed in Part III, Chapter 10).

A venue that increases the likelihood of State responsibility already *de lege lata* in this regard, or at least mitigates the gap, is through the duty to take precaution that requires a lower threshold of fault than international criminal provisions (Amoroso, 2020, pp. 149–150).[30] The obligation to "do everything feasible" might be violated through negligence. Responsibility due to the breach of due diligence will be the focus of Chapter 7.4.

4.3 AWS and Damage

According to the doctrine (Graefrath, 1984, pp. 34–36)[31] and the ARSIWA, damage is not an essential element of the internationally wrongful act. As is widely known, the two conditions for an internationally wrongful act are a breach of an international obligation, and conduct attributable to the State.

> "If there is no internationally wrongful act so long as the event has not occurred, the reason is that until then the State's conduct has not resulted in the breach of an international obligation. It is really the objective element of the internationally wrongful act that is missing. In other words, the occurrence of an external event is a condition for the breach of an

---

[30] Amoroso affirms that "(...)already *de lege lata* – the failure to take proper risk mitigation measures resulting in civilian losses (or otherwise unlawful deprivations of life) is likely to trigger State responsibility, even when it is not possible to establish the direct criminal responsibility of AWS's operators. Furthermore, the application of due diligence standard would make it possible to hold a State responsible in "hard" many hands scenarios (...)" (Amoroso, 2020, p. 150).

[31] Graefrath states that "Contrary to former opinions, in contemporary international law, damage is no longer considered to be a constituent element of an internationally wrongful act entailing State responsibility. The violation of an international obligation that can be attributed to a State is sufficient to establish its international responsibility. After a short discussion, this standpoint was generally accepted in the International Law Commission. Naturally, there are numerous rules where, as a premise for violation of an obligation, causing of material damage is expressly required. In such cases a violation of an obligation can be in question only if material damage was caused, since only then does a violation of the obligation exist (...)
This was clearly formulated by Ago and the International Law Commission shared his position : Committing an international wrongful act means committing . . . a breach of an obligation of international law towards another State, and not inflicting damage on that State. The act is wrongful even if there is no damage, particularly damage of a financial nature . . . The State must take reparation for the delinquency itself, for the disturbance caused in international legal relations by the breach of its own legal obligations towards another State, and not the 'damage' which that delinquency may have caused . . . it is the delinquency for which reparation is made and not the damage which may result there from" (Graefrath, 1984, pp. 34–36).

international obligation, and not a new element which has to be combined with the breach for there to be a wrongful act" (UN ILC, 1973, p. 183).

The ILC commentaries to the ARSIWA state that the primary rule establishes the necessity or not of damage (UN ILC, 2001, p. 36).[32] In the case of AWS, there is no specific rule on the issue, and IHL rules such as distinction, proportionality, and the IHRL prohibition on arbitrary deprivation of life (UN ICCPRC, 2019)[33] do not require damage.

The question we pose in this section is: can a breach of an international obligation perpetrated through AWS conduct occur without damage?

Despite being hard to externalize and prove the violation of an autonomous device before the damage occurs, AWS might be programmed for unlawful killings, and damage is not required. For instance, if AWS are programmed to conduct indiscriminate attacks and the AWS is put in the field, there is a violation even before the indiscriminate attack occurs. Nonetheless, except for those "easy cases," considering the inherent characteristics of AWS and their action chain presents a degree of unpredictability, it might be difficult to prove State responsibility without damage.

In short, despite being unnecessary, the damage is a factor that boosts State international responsibility for "making the breach visible." Furthermore, once the damage is ascertained, States are likely to be more prone to take responsibility for their breaches (Mauri, personal communication, July 5, 2022).[34]

---

32 According to ILC, "(9) Thus there is no exception to the principle stated in article 2 that there are two necessary conditions for an internationally wrongful act—conduct attributable to the State under international law and the breach by that conduct of an international obligation of the State. The question is whether those two necessary conditions are also sufficient. It is sometimes said that international responsibility is not engaged by conduct of a State in disregard of its obligations unless some further element exists, in particular, "damage" to another State. But whether such elements are required depends on the content of the primary obligation, and there is no general rule in this respect" (UN ILC, 2001b, p. 36).

33 The General Comment No. 36 on Article 6: right to life states that "3. The right to life is a right that should not be interpreted narrowly. It concerns the entitlement of individuals to be free from acts and omissions that are intended or may be expected to cause their unnatural or premature death, as well as to enjoy a life with dignity. Article 6 of the Covenant guarantees this right for all human beings, without distinction of any kind, including for persons suspected or convicted of even the most serious crimes.

4. Paragraph 1 of article 6 of the Covenant provides that no one shall be arbitrarily deprived of life and that this right shall be protected by law. It lays the foundation for the obligation of States parties to respect and ensure the right to life, to give effect to it through legislative and other measures, and to provide effective remedies and reparation to all victims of violations of the right to life" (UN ICCPRC, 2019).

34 This is a citation from the author´s Ph.D. dissertation. During an interview for the author´s Ph.D., Diego Mauri, in his personal capacity, stated that: "Empirically damage is a boosting factor for international responsibility, even if it is not, as it is known, a component of international responsibility (…) Once you have damage, and this ascertained and objective, then probably States will be more likely to adopt new frameworks of responsibility such as strict liability" (Mauri, personal communication, July 5, 2022).

4.4 AWS and Challenges to State the Time of an Internationally Wrongful Act

Article 14 of the ARSIWA (UN ILC, 2001c, art. 14) deals with the time of the breach of an international obligation. It occurs at the moment when the act is performed, even if the effects are protracted in time. Article 14 further states that if the act has a continuous character, the breach extends over the period in which the act remains in non conformity with the international obligation. If the obligation is to prevent,[35] the breach arises when the event occurs and remains over the period the situation is in non conformity with the international obligation (UN ILC, 2001b, p. 62).[36] As stated in the ILC comments, the occurrence and duration of an internationally wrongful act depend on the primary obligation and the specific factual situation; article 14 offers a basic framework (UN ILC, 2001b, p. 59).

Determining the time of a violation caused by AWS has an additional layer of challenges. Unlike other wrongful acts, actions that trigger the breaches probably will not be taken by a human being but through AI. When one takes a wider perspective, there will always be some human action, but humans might not have been capable of foreseeing the course of actions taken by AWS. There might also be a vast geographical and temporal gap between the human decision to putan AWS in the field and the outcome of this decision. This detachment from the use of force in the context of AWS is even greater than with other technologies, such as drones and fire-and-forget missiles (Crootof, 2018, pp. 64–67), as the decision is taken without the necessity of human intervention, and algorithms might change over time due to machine learning. Moreover, it is possible that despite selecting an unlawful target, a human agent or the machine itself can prevent damage.

35 Does the principle of prevention require that it is always possible to abort the mission of AWS? This falls into the category of primary obligation and falls outside the scope of the present work. Nonetheless, it is an issue that needs to be further discussed.
It is important to note that the obligation on prevention stated in article 36 of Additional Protocol I of the 1949 Geneva Conventions, which foresees weapon review, is extremely relevant for AWS.

36 According to ILC "(14) Paragraph 3 of article 14 deals with the temporal dimensions of a particular category of breaches of international obligations, namely the breach of obligations to prevent the occurrence of a given event. Obligations of prevention are usually construed as best efforts obligations, requiring States to take all reasonable or necessary measures to prevent a given event from occurring, but without warranting that the event will not occur. The breach of an obligation of prevention may well be a continuing wrongful act, although, as for other continuing wrongful acts, the effect of article 13 is that the breach only continues if the State is bound by the obligation for the period during which the event continues and remains not in conformity with what is required by the obligation. For example, the obligation to prevent trans boundary damage by air pollution, dealt with in the Trail Smelter arbitration was breached for as long as the pollution continued to be emitted. Indeed, in such cases the breach may be progressively aggravated by the failure to suppress it. However, not all obligations directed to preventing an act from occurring will be of this kind. If the obligation in question was only concerned to prevent the happening of the event in the first place (as distinct from its continuation), there will be no continuing wrongful act. If the obligation in question has ceased, any continuing conduct by definition ceases to be wrongful at that time" (UN ILC, 2001b, p. 62).

The question in this section is when does a breach by AWS occur? Which acts should be considered? The moment of programming? The moment the device with machine learning reprograms unlawfully? The moment AWS became part of the operation or the moment in which damage occurred? Furthermore, if there is timely deactivation, has a wrongful act occurred or not? The ARSIWA does not provide a basic framework. Furthermore, existing primary rules do not answer the challenges AWS pose to international responsibility.

With AWS-related breaches, programming or "(...) many of those last human interventions—including the development of an IHL compliant program instructing the autonomous weapon when to use lethal force—may occur before an armed conflict exists" (Sassòli, 2014, p. 325). *Tempus comissi delicti* is a determinant factor in assessing the applicable law. In international criminal law, war crimes can only occur in the context of armed conflicts. IHL contemplates "(...) all conduct of a State aimed at having effects during an armed conflict" (Sassòli, 2014, p. 308). The complex issue will be to determine whether specific programming does or does not aim to have effects in an armed conflict. Must this intention be regarding a specific armed conflict or armed conflicts in general? Sassòli argues that a State that uses a weapon programmed not to comply with IHL before the armed conflict breaches the duty of precaution (Sassòli, 2014). We agree with Sassòli but think he dealt only with the tip of the iceberg, the easy case in which it is demonstrated that a weapon was programmed to violate humanitarian law. What if the State organs do not know about the harmful programming? Does international humanitarian law apply?

In conclusion, AI adds a challenging roadblock in stating the *tempus comissi delicti,* and many times, the wrongful act will only be apparent after damage occurs. As discussed in the previous section, damage is not an essential element of the wrongful act and depends on the primary rule, but it might be challenging to prove AWS violation before the damage.

Further challenging is that the action can be composite and, under article 15 of ARSIWA (UN ILC, 2001b, pp. 62–63),[37] a breach of international obligations through a series of acts or omissions occurs when those taken together are sufficient to constitute a wrongful act. Composite acts are, for instance, the case

---

[37] According to ILC's commentaries: "(...) article 15 deals with a further refinement, viz. the notion of a composite wrongful act. Composite acts give rise to continuing breaches, which extend in time from the first of the actions or omissions in the series of acts making up the wrongful conduct.

(...)

(2) Composite acts covered by article 15 are limited to breaches of obligations which concern some aggregate of conduct and not individual acts as such.

(4) It is necessary to distinguish composite obligations from simple obligations breached by a "composite" act. Composite acts may be more likely to give rise to continuing breaches, but simple acts can cause continuing breaches as well. The position is different, however, where the obligation itself is defined in terms of the cumulative character of the conduct, i.e., where the cumulative conduct constitutes the essence of the wrongful act. Thus, apartheid is different in kind from individual acts of racial discrimination, and genocide is different in kind from individual acts even of ethnically or racially motivated killing" (UN ILC, 2001b, pp. 62–63).

of swarms that will be discussed in section 5.3. Each swarm member is an AI agent that might act in unpredictable ways. Thus, it will be hard to know when it suffices for the violation to occur and set the breach's time before the occurrence of damage. The probationary field will be very complex or even unsolvable. A further challenge is added if the composite acts stated in article 15 are attributable to more than one State since single acts could be per se lawful but jointly considered unlawful. The ARSIWA, as regards *tempus comissi delicti,* probably presumed that all acts would be attributable to one State (Gattini, 2014, p. 23),[38] which might not be the case on the ground.

Finally, also regarding the time frame, it is significant to note that the 2012 and 2023 US DOD Directive 3009 deal with the time issue[39], and at the CCW GGE, the US delegation stressed the importance of a temporal limitation:

> "(…) even an attack against authorized targets could be "unintended" if there are significant changes to the factual context between the time of authorization and the engagement (for example, if a cease-fire agreement is negotiated)." In this regard, DoD Directive 3000.09 requires that autonomous and semi-autonomous weapon systems be designed to "[c] complete engagements in a timeframe consistent with commander and operator intentions and, if unable to do so, to terminate engagements or seek additional human operator input before continuing the engagement" (US Delegation 2018a, p. 2).

A time frame limitation is a tool to help assess the *tempus comissi delicti* and reduce causalities (Margulies, 2016, p. 22)[40] and help ensure responsibility, as will be further discussed in Part III.

---

38 Gattini states that "Indeed, there is no hint in the ILC Commentary that the single act or omission forming the composite wrongful act could be per se lawful. The ILC's neglect of this specific issue could be explained by the fact that at that point its attention was focused on the issue of the *tempus delicti commissi* and that it possibly took for granted that the entire sequence of conduct would be attributable to a single state" (Gattini, 2014, p. 23).

39 For comments on the 2023 US Directive 3009 see Appendix N.

40 Margulies proposes that "To lower the probability of an AWS going rogue, each AWS should include time and distance defaults. The AWS should be programmed to move into a default hibernation mode after a relatively short, discrete period, which might be 24-96 hours. During this period, a human with remote access to the AWS's software could override the default and authorize continued operation for an additional increment of time" (Margulies, 2016, p. 22).

CHAPTER

5

The Internationally Wrongful Act of a State: Responsibility of a State for AWS Violations in Connection with the Act of Another State

Situations might occur in which multiple States are involved in a breach of international law, and the ARSIWA deals with some of them(UN ILC, 2001b, p. 124).[1] ILC commentaries stress that the guiding principle is independent responsibility, meaning each State bears its obligations and corresponding responsibility.[2] For example, an internationally wrongful act might result from the collaboration of States: a) by independent action, b) through a common organ, or c) when a State acts on behalf of another State (UN ILC, 2001b, p. 64). There are even situations in which one State is responsible for the internationally wrongful act of another.

> "In most cases of collaborative conduct by States, responsibility for the wrongful act will be determined according to the principle of independent responsibility (...) But there may be cases where conduct of the organ of one State, not acting as an organ or agent of another State, is nonetheless

---

1 As mentioned in the methodology, this thesis mainly follows the structure of the ARSIWA. Thus, this section will deal with the situation of multiple States involved in a breach of an international obligation, addressed by Chapter IV of the ARSIWA. Nonetheless, we will also analyze articles outside the framework of Chapter IV, such as articles 6 and 47 of the ARSIWA, since they also deal with the involvement of more than one State in the breach.

2 ILC´s commentaries state that"(3) It is important not to assume that internal law concepts and rules in this field can be applied directly to international law. Terms such as "joint", "joint and several" and "solidary" responsibility derive from different legal traditions and analogies must be applied with care. In international law, the general principle in the case of a plurality of responsible States is that each State is separately responsible for conduct attributable to it in the sense of Article 2. The principle of independent responsibility reflects the position under general international law, in the absence of agreement to the contrary between the States concerned. In the application of that principle, however, the situation can arise where a single course of conduct is at the same time attributable to several States and is internationally wrongful for each of them. It is to such cases that article 47 is addressed" (UN ILC, 2001b, p. 124).

chargeable to the latter State, and this may be so even though the wrongfulness of the conduct lies, or at any rate primarily lies, in a breach of the international obligations of the former" (UN ILC, 2001b, p. 64).

It is also possible that the same conduct is attributable to more than one State, as stated by article 47, in which case the responsibility of each of those States can be invoked (UN ILC, 2001c, art. 47).

Situations of shared responsibility are challenging to State responsibility (SHARES Project, n.d.),[3] and AWS add an additional layer of challenges. This chapter will focus on some specific challenges AWS might pose to situations of shared responsibility, namely: Obligation not to aid or give assistance to another State and AWS; AWS designed, manufactured, and programmed by one State and deployed by another; and swarms of AWS composed of weapons of different States.

This chapter will set aside the problem of a human action paradigm of attribution discussed in Chapter 4, and the focus will be on dealing with multiple States involved in the breach.

5.1 Obligation not to Aid Assistance to another State: the Problem of Intent

According to article 16 of the ARSIWA, a State that aids or assists another State in the commission of an internationally wrongful act of that other State bears responsibility.[4] It sets forth the ancillary responsibility for facilitating the assisted State to violate international law and does not address the assisted State's responsibility (ILC, 2002c, art. 16).[5]

First, it is necessary to emphasize that article 16 is not a self-standing primary norm but a secondary norm of attribution of responsibility. Thus, for example, for genocide it is still necessary that the person aiding assistance acts knowing the *dolus speciallis,* which characterizes the crime of genocide (ICJ, 2007, par. 420–421).[6]

---

3 The SHARES project, led by Professor André Nollkaemper, studied the challenges of shared responsibility in depth. (SHARES Project, n.d.).

4 International Law Commission, Draft Articles on Responsibility of States for Internationally Wrongful Acts, 2001, Supplement No. 10 (A/56/10), chp.IV.E.1 Article 16. "Aid or assistance in the commission of an internationally wrongful act
A State that aids or assists another State in the commission of an internationally wrongful act by the latter is internationally responsible for doing so if: (a) that State does so with knowledge of the circumstances of the internationally wrongful act; and (b) the act would be internationally wrongful if committed by that State."

5 ICJ in the Bosnian Genocide case recognized that article 16 is part of customary international law (ICJ, 2007, par. 420).

6 ICJ stated that "The Court sees no reason to make any distinction of substance between "complicity in genocide", within the meaning of Article III, paragraph (e), of the Convention, and the "aid or assistance" of a State in the commission of a wrongful act by another State

(*Contd.*)

Back to article 16, ILC commentaries foresee three requirements for the responsibility for aiding assistance can attach: knowledge of the circumstances of the wrongful act, aims at facilitating the commission of the act, and the act would also be wrongful if committed by the State itself. Material and financial assistance does not usually mean aiding assistance to commission internationally wrongful acts. The ILC explains that aiming at facilitating requires a specific subjective element: that the State organ intended the course of conduct.

> "A State is not responsible for aid or assistance under article 16 unless the relevant State organ intended, by the aid or assistance given, to facilitate the occurrence of the wrongful conduct and the internationally wrongful conduct is actually committed by the aided or assisted State" (UN ILC, 2001b, p. 66).

The necessity of intention of the State organ may be a relevant bar to responsibility in the context of AWS. Many challenges to individual responsibility in the context of AWS breaches, arising from their unpredictability, came into play. For responsibility to ensure under article 16, an organ of the assisting state must have aimed that AWS breach international law.

If this high threshold of intent is satisfied, which means that the assisting State has knowledge about the circumstances of the breaches by another State and provides AWS aiming at supporting that breach, which would also be a violation if committed by the programmer State, it is responsible for aiding and assisting the commission of the wrongful act. What if the aiding State provides the AWS without an intention to facilitate the breach of an international obligation?

Sassòli highlights the special humanitarian rule in this regard and argues that a State is responsible if it provides weapons for a State that systematically commits violations of international humanitarian law with this kind of weapon, even if they could be used lawfully. He states, "Such a strict standard may not be that of the ILC in its Commentary" but is rooted in the special obligation to refrain from assisting in violations and also to "ensure respect" for humanitarian law (Sassòli, 2002, p. 413).

Thus, in the context of armed conflicts, a State that provides AWS or programs those weapons to a State that systematically breaches international law using them

---

within the meaning of the aforementioned Article 16—setting aside the hypothesis of the issue of instructions or directions or the exercise of effective control, the effects of which, in the law of international responsibility, extend beyond complicity. In other words, to ascertain whether the Respondent is responsible for "complicity in genocide" within the meaning of Article III, paragraph (e), which is what the Court now has to do, it must examine whether organs of the respondent State, or persons acting on its instructions or under its direction or effective control, furnished "aid or assistance" in the commission of the genocide in Srebrenica, in a sense not significantly different from that of those concepts in the general law of international responsibility" (ICJ, 2007, par. 420).

"(...)there is no doubt that the conduct of an organ or a person furnishing aid or assistance to a perpetrator of the crime of genocide cannot be treated as complicity in genocide unless at the least that organ or person acted knowingly, that is to say, in particular, was aware of the specific intent (*dolus specialis*) of the principal perpetrator (...)" (ICJ, 2007, par. 421).

is responsible if it is aware of systematic violations, even if it does not meet the higher threshold of article 16 of the ARSIWA.

Nonetheless, responsibility for aiding or assisting might be challenging outside the context of armed conflicts, considering that article 16 requires intent from the State agent, an element that will be hard to demonstrate in the case of AWS breaches. Also, despite the lower threshold, responsibility for aid or assistance might be challenging in the context of IHL, as it is necessary to demonstrate awareness of systematic violations.

5.2 AWS Placed at the Disposal of a State by Another State: AWS Designed, Manufactured, and Programmed by one State and Deployed by Another

Article 6 of the ARSIWA foresees that "conduct of an organ placed at the disposal of a State by another State shall be considered an act of the former State under international law if the organ is acting in the exercise of elements of the governmental authority of the State at whose disposal it is placed" (UN ILC, 2001c, art. 6).

First, we acknowledge that article 6 is under the ARSIWA Chapter II, which deals with Attribution of a Conduct to a State, and not in Chapter IV, which deals with Responsibility of a State in Connection with the Act of Another State. However, considering that it addresses multiple States involved in the breach of international obligation, which is the object of this chapter, we opted to place it in this section.

As previously discussed, the notion of "organ" is based on a human action paradigm and does not apply to AWS. Nonetheless, to make possible the analysis, consider for this section that the human action paradigm is overruled by a specific primary rule or by a paradigm shift in their comprehension and that the term 'organ' embraces AWS.

To fit in the framework of article 6, AWS must be an organ of the sending State, cannot be from the private sector, and must be under the authority of the receiving State (UN ILC, 2001b, p. 44).[7] What would it mean to have authority?

[7] ILC commentaries state that"(5)There are two further criteria that must be met for article 6 to apply. First, the organ in question must possess the status of an organ of the sending State; and secondly, its conduct must involve the exercise of elements of the governmental authority of the receiving State. The first of these conditions excludes from the ambit of Article 6 the conduct of private entities or individuals which have never had the status of an organ of the sending State. For example, experts or advisers placed at the disposal of a State under technical assistance programs do not usually have the status of organs of the sending State. The second condition is that the organ placed at the disposal of a State by another State must be "acting in the exercise of elements of the governmental authority" of the receiving State. There will only be an act attributable to the receiving State where the conduct of the loaned organ involves the exercise of the governmental authority of that State. By comparison with the number of cases of cooperative action by States in fields such as mutual defence, aid, and development, article 6 covers only a

(*Contd.*)

Does authority mean having written the program? Does it mean to deploy the AWS? Or does authority mean to be able to abort AWS missions?

The question of this section is whether AWS designed, manufactured, and programmed by one State and deployed by another can be considered to be in the exercise of the latter's authority.

ILC commentaries state that article 6 of the ARSIWA deals not with ordinary cooperation but with the situation in which an organ acts under the exclusive direction and control of the receiving State and does not receive the sending State's instructions. Mere aid or assistance does not fall under article 6 (UN ILC, 2001b, p. 44).[8]

> "The notion of an organ "placed at the disposal of" the receiving State is a specialized one, implying that the organ is acting with the consent, under the authority of and for the purposes of the receiving State. Not only must the organ be appointed to perform functions appertaining to the State at whose disposal it is placed, but in performing the functions entrusted to it by the beneficiary State, the organ must also act in conjunction with the machinery of that State and under its exclusive direction and control, rather than on instructions from the sending State. Thus article 6 is not concerned with ordinary situations of inter-State cooperation or collaboration, pursuant to treaty or otherwise" (UN ILC, 2001b, p. 44).

Does programming mean instructions of the sending State? Is programming sufficient to exclude exclusive control?

We argue that even if international law evolves to consider AWS as State organs, it is relevant to note that *de lege lata*, AWS programmed by the sending State, cannot be deemed under the authority of the receiving State. First, programming AWS is a complex operation that might involve many "hands" and might not be fully understandable even to the programmers and, thus, will not likely be under the exclusive direction and control of the receiving State. One could argue that, like humans might be under the direction and control of a State and behave in unexpected ways, AWS might also do so. Nonetheless, we claim that AWS differ from humans in this regard since, as previously explored, AWS present a degree of unpredictability and outpace the speed of decision-making of humans, making it harder to exercise control. *De lege ferenda*, a statement

specific and limited notion of "transferred responsibility". Yet, in State practice the situation is not unknown" (UN ILC, 2001b, p. 44).

8 ILC commentaries state that"On the other hand, mere aid or assistance offered by organs of one State to another on the territory of the latter is not covered by article 6. For example, armed forces may be sent to assist another State in the exercise of the right of collective self-defence or for other purposes. Where the forces in question remain under the authority of the sending State, they exercise elements of the governmental authority of that State and not of the receiving State. Situations can also arise where the organ of one State acts on the joint instructions of its own and another State, or there may be a single entity which is a joint organ of several States. In these cases, the conduct in question is attributable to both States under other articles of this chapter" (UN ILC, 2001b, p. 44).

regarding the application (or not) of Article 6 to AWS is necessary to elucidate the situation. Another option is to resort to an interpretive extension, by analogy, of the scope of Article 6, however, as it will be shown in section 8.2.1.1. "Analogical interpretation possible framework in the context of AWS," would be too-far a stretch for current international law.

5.3 Swarms Composed of AWS of Different States

Swarms of AWS, or cooperative autonomy, is a collectivity of AWS working together like a colony of bees. They can happen in the air, on land, or at sea (Scharre, 2018)[9] and are part of evolving warfare (Scharre, 2018, p. 17). Swarms will challenge multilateral debate and negotiations on AWS (Ekelhof and Persi, 2020).

We acknowledge the viewpoint that considering that AWS are systems, and systems are a set of connected things or devices that operate together (Cambridge Dictionary, n.d.), as discussed in the proposed definition of AWS, one could envision that the whole swarm is one AWS (Crootof, personal communication, May 26, 2022).[10] However, we claim that each device of the swarm that meets the definition of AWS should be considered as one AWS and the collectivity of them a swarm.

The question is which State will be responsible if a swarm of AWS composed of devices belonging to two or more States breaches international law.

We agree with the view that already *de lege lata*, "if a swarm was in operation, even if you had different countries with weapons incorporated in the swarm, there has to be one country responsible for the swarm, commanding the swarm" (Scharre, personal communication, May 27, 2022). However, if there is

9 Paul Scharre exemplifies swarms of drones. "In 2016, the United States demonstrated 103 aerial drones flying together in a swarm that DoD officials described as "a collective organism, sharing one distributed brain for decision-making and adapting to each other like swarms in nature" (Not to be outdone, a few months later China demonstrated a 119-drone swarm). Flying together, a drone swarm could be far more effective than the same number of drones flying individually" (Scharre, 2018, p. 21).
Scharre also exemplifies swarms of boats that were not AWS: "Swarms aren't merely limited to the air. In August 2014, the U. S. Navy office of Naval Research (ONR) demonstrated a swarm of small boats on the James River in Virginia by simulating a mock strait transit in which the boats protected a high-value Navy Ship against possible threats, escorting it through a simulated high-danger area" (Scharre, 2018, p. 21). The swarms of drones and boats mentioned were not AWS, but the next development step is likely to be swarms of AWS. See also (Crootof, 2018, pp. 80–81).

10 This is a citation from the author's Ph.D. dissertation. During an interview for the author's Ph.D., Rebecca Crootof, in her personal capacity, stated that: "It depends on whether or not you conceive of AWS as an individual, discrete entities, or if you conceive AWS as systems because, under an expensive definition, the swarm is as a whole the AWS. A system can be an AWS without each individual, independent component being an AWS. The thing as a whole could be conceptualized as the AWS. (...)But we tend to think about AWS as being a robotic soldier, as opposed to thinking about them as a network of sensors and drones and landmines" (Crootof, personal communication, May 26, 2022).

no previously agreed or established State responsible or if it is not proved which State was responsible, States still need to be held accountable. Therefore, the following responsibility framework is proposed depending on the kind of swarm.

There are four kinds of swarm command and control models: centralized coordination, hierarchical coordination, consensus-based coordination, and emergent coordination (Scharre, 2018, p. 17).

First, we will analyze centralized coordinated and hierarchical coordinated swarms and the responsibility of the State for AWS breaches. In those command-and-control patterns, it is possible to apply the principle of specific responsibility, which means that each State bears its own obligations and corresponding responsibility. Articles 16 (aiding assistance) and 17(direction and control over the commission of an internationally wrongful act) might apply depending on the situation.

In centralized coordination, even if the swarm is composed of AWS from more than one State, a single AWS controller receives information from all the individual swarm elements and then instructs them (Scharre, 2018, p. 19). Thus, the primary responsibility for breaches will be the State controlling the actions, even if this central controller is an AWS exerting this control through AI. Following the principle of specific responsibility, the State responsible for the AWS commanded by the other State will be responsible to the extent they contributed to the breach.

The Swarm´s hierarchical coordination means the AWS that compose it work in a chain of command and control of top-down orders, similar to a military organization hierarchy (Scharre, 2018, p. 19). Therefore, if the swarm commits the breach, one must analyze the chain of command to assess which State is responsible for the violation. It will be necessary to examine the concrete situations to find out the "subordinated AWS" and the "commanding AWS" to analyze the degree of responsibility of each State.

Significantly enough, it will be difficult for the claimant to prove who is responsible for both types of command and control, namely centralized coordination, and hierarchical coordination. A strict liability regime or reversing the burden of proof would be most helpful.

Decentralized command and control models are consensus-based coordination and emergent coordination. Unlike the two models above, there is no central or hierarchical decision-making.

Consensus-based coordination is a decentralized swarm control and command model in which the compounding devices collectively decide. All AWS that are part of the swarm "communicate with one another simultaneously and collectively decide on a course of action. They could do this by using "voting" or "auction" algorithms to coordinate behavior" (Scharre, 2018, p. 19).

If a breach occurs and the swarm is composed of devices of more than one State operating under consensus-based decision-making, both States will be responsible under article 47 of the ARSIWA. In those cases, applying the principle of specific responsibility is almost impossible. Even if the AWS of a State did not "vote" for the breach, they agreed on the decision-making process, which *per se* makes all of them responsible. Thus, both States will be responsible for the breach. Nonetheless, in the concrete case, it might be necessary to analyze if, for instance,there were far more AWS from one State than another and if the contribution to the breach was minor. In such a situation, there might be a proportional distribution of responsibility.

Finally, the most decentralized command and control model is emergent coordination, in which each AWS makes its own decisions based on the other AWS of the swarm. Note that "Simple rules for individual behavior can lead to very complex collective action, allowing the swarm to exhibit "collective intelligence" (Scharre, 2018, pp. 19–20). This model "is how flocks of birds, colonies of insects, and mobs of people work, with coordinated action arising naturally from each individual making decisions based on those nearby" (Scharre, 2018, pp. 19–20).

> You could have emergent behavior where a swarm takes some action, and then the person that put the swarm into operation is surprised by what it does. The point of the swarm is not to have direct human control over the actions but to delegate those decisions to the swarming agents themselves and allow them to determine what actions to take. So you could have certain circumstances where a person is surprised and then says, well, I didn't know it was going to do that (Scharre, personal communication, May 27, 2022).

If a breach occurs and the swarm composed by AWS of more than one State is operating under emergent coordination decision-making, both States will be responsible, under article 47 of ARSIWA, it being again hard to apply the principle of specific responsibility. If States deploy their AWS in such a swarm, they agree on the decision-making model of "collective intelligence" which makes it hard to dissociate every component of the swarm's share of responsibility.

The proposed allocations of responsibility for the four kinds of swarms were based on an interpretation of the principle of individual responsibility. Nonetheless, the issue is unclear and might interpreted differently, especially in consensus-based and emergent coordination cases. Determining and proving the extent of responsibility of each State in the case of swarms would be extremely hard.

CHAPTER

6

The Internationally Wrongful Act of a State: Force Majeure; and the Content of the International Responsibility of a State: Assurances of Non-Repetition

According to the ARSIWA, consent, self-defense, force majeure, distress, and necessity are the circumstances precluding wrongfulness. This chapter will only analyze force majeure as it deals with unforeseen events, and, thus, implies specific challenges to AWS misdoings. This chapter will set aside the problem of a human-centered paradigm of attribution addressed in Chapter 4, and the focus will be on force majeure as a circumstance precluding wrongfulness.

Next, this chapter examines Part II of ARSIWA regarding the general principles on the content of international responsibility. As previously mentioned, we will not analyze all provisions but only assurance of non-repetition, as AWS pose extra challenges to it. Therefore, for instance, we will not deal with Chapter II of Part II, "Reparation of Injury," as, in our view, there are no fundamental differences in reparation when breaches occur in the context of AWS.

6.1 The Danger of Excluding Responsibility Based on Force Majeure

The question posed in this section is if AWS actions, due to their inherent degree of unpredictability, may fall within the concept of *force majeure*, thus precluding wrongfulness.

Note that the question is not whether *force majeure* applies to violations committed by AWS, since there might be situations, for instance, natural disasters such as an unforeseen earthquake that can unexpectedly activate AWS. In this sense, we agree that *force majeure* and all other circumstances precluding wrongfulness are fully applicable to States that deploy AWS, if the required elements are met (Ford, 2017, pp. 475–477).

It is necessary to analyze if AWS actions may come within the *force majeure* concept, thus precluding wrongfulness, for the mere fact of being opaque and unpredictable. In other words, "increasing autonomous capabilities in targeting will make it tempting for States to plead force majeure to evade international responsibility for unforeseen and unforeseeable 'decisions' taken by the autonomous system" (Mauri, 2022, p. 2018). The concern is that *force majeure* becomes a common defense for States that commit internationally wrongful acts in the context of AWS, thus creating a relevant obstacle to the international responsibility of States.

Article 23 of the ARSIWA states that *force majeure*, an irresistible force, or an unforeseen event, precludes wrongfulness. It embraces natural and human-created events which include unexpected AWS wrong doings (Mauri, 2022, p. 2018). The ARSIWA requires that the event is outside the control of the State and that, in the circumstances, the State cannot perform the obligation.

A defendant State could claim that AWS actions fall into *force majeure* since AWS act through AI, and their actions are inherently unpredictable, even to the programmers. It is also possible that the defendant affirms that the breach was out of its control since the pace of decision-making of AI devices far outpace the speed of human decision-making or due to machine learning algorithms (UN ILC, 2001b, p. 76).[1] They might argue that there was no element of free choice of the State (UN ILC, 2001b, p. 76),[2] not concerning the AWS deployment, but regarding the course of action it might take. Thus, actions by AWS might be deemed unforeseen events.

Nonetheless, under the ARSIWA *force majeure* does not apply if: (a) the State's conduct led to or contributed to the situation of *force majeure*, or (b) the State assumed the risk of the situation of *force majeure* (UN ILC, 2001c, art. 23).

The first hypothesis of inapplicability of *force majeure* is if "(...) the situation (...) is due, either alone or in combination with other factors, to the conduct of the State invoking it." A State that deploys AWS almost undoubtedly contributes with its conduct to *force majeure*, as it deployed an inherently unpredictable weapon. However, if the contribution that led to the unforeseen event was unwilling and in good faith, States can claim *force majeure* (Crootof, personal communication, May 26, 2022).

The ILC commentaries highlight that the risk assumption "(...) must be unequivocal" (UN ILC, 2001b, p. 78, par. 10). It refers to cases where States

---

1 ILC commentaries state that "(2) A situation of force majeure precluding wrongfulness only arises where three elements are met: (a) the act in question must be brought about by an irresistible force or an unforeseen event; (b) which is beyond the control of the State concerned; and (c) which makes it materially impossible in the circumstances to perform the obligation. The adjective "irresistible" qualifying the word "force" emphasizes that there must be a constraint which the State was unable to avoid or oppose by its own means. To have been "unforeseen" the event must have been neither foreseen nor of an easily foreseeable kind. Further the "irresistible force" or "unforeseen event" must be causally linked to the situation of material impossibility, as indicated by the words "due to force majeure ... making it materially impossible." Subject to paragraph 2, where these elements are met, the wrongfulness of the State's conduct is precluded for so long as the situation of force majeure subsists" (UN ILC, 2001b, p. 76).

2 ILC commentaries state that "Force majeure differs from a situation of distress (art. 24) or necessity (art. 25) because the conduct of the State which would otherwise be internationally wrongful is involuntary or at least involves no element of free choice" (UN ILC, 2001b, p. 76).

expressly accepted, mainly by way of agreement, to carry the risk of a breach (e.g., by ratifying a treaty establishing a strict liability regime). The ILC also requires that the risk assumption is "directed towards those to whom the obligation is owed" (UN ILC, 2001b, p. 78). The deployment of AWS will hardly mean unequivocally accepting the risk of occurrence of *force majeure* if the only unforeseen event is the AWS. We stated, "will hardly mean," as if such an assumption was unequivocal or not, depends on the concrete situation analyses of express risk assumption, and probably almost no State will assume or agree it assumed the risk (Crootof, personal communication, May 26, 2022).

In sum, our perspective is that, in general, the mere fact that AWS committed the violation should not allow the defense argument of *force majeure*. The reason is that the deploying State contributed to the breach. However, if the contribution was in good faith and unwilling, States can claim *force majeure* (Crootof, personal communication, May 26, 2022).[3] Furthermore, only in very few cases will it be proven that States expressly and unequivocally assumed the risk of *force majeure.* Considering the previously stated good faith defense, the probationary challenge of demonstrating its inapplicability on a case-by-case basis, and that the ARSIWA might be interpreted in a different venue, *de lege ferenda*, a specific provision that *force majeure* does not apply solely based on the inherent unpredictability of AWS is advisable. *Force majeure* inapplicability will be further discussed in Part III.

It is relevant to note that article 26 of the ARSIWA states that *force majeure* does not preclude the wrongfulness of acts that violate peremptory norms of international law UN ILC 2001c, art. 26).[4] Within the category of peremptory norms are essential principles and rules of IHL, including the distinction, proportionality, and the IHRL prohibition of arbitrary deprivations of life (Amoroso, 2020, p. 147). The ILC's Draft Conclusion 23 (UN ILC, 2019)[5] states that basic rules of international humanitarian law are peremptory norms. The ICJ and the ICTY also stated that basic rules of international humanitarian law

3 This is a citation from the author's Ph.D. dissertation. During an interview for the author's Ph.D., Rebecca Crootof, in her personal capacity, stated that: "States can still claim *force majeure* – that an act isn't wrongful due to an irresistible force or unforeseen event that makes it impossible for the State to perform its obligation. This defense only applies if the situation is not due to the acts of the State invoking the defense or if the State has not assumed the risk of the situation occurring. But those limitations often won't apply. First, States can claim this defense when the State has unwittingly contributed to the occurrence of the material impossibility, so long as its act was done in good faith and the act didn't make the bad event any more foreseeable. And second, you only get to say that the State has assumed the risk if the states unequivocally accept the risk of accidental harm. No State is going to say that it has done that. So, in many situations, States will still get to invoke *force majeure*" (Crootof, personal communication, May 26, 2022).

4 In this regard, ILC comments that "(4) It is, however, desirable to make it clear that the circumstances precluding wrongfulness in chapter V of Part One do not authorize or excuse any derogation from a peremptory norm of general international law" (UN ILC, 2001b, p. 85, par 4).

5 United Nations International Law Commission, Draft Conclusions on Peremptory Norms of General International Law (Jus Cogens), at Conclusion 23 states that "Non-exhaustive list without prejudice to the existence or subsequent emergence of other peremptory norms of general international law (jus cogens), a non-exhaustive list of norms that the International Law Commission has previously referred to as having that status is to be found in the annex to the present draft conclusions (…)
(d) The basic rules of international humanitarian law" (UN ILC, 2019, par. 57).

are peremptory, and some scholars affirm that all rules of humanitarian law are peremptory (Sassòli, 2002, p. 414).

> At least from the standpoint of the concept of *jus cogens* under the law of treaties, international humanitarian law itself supports this view when it prohibits separate agreements that adversely affect the situation of protected persons. It would be difficult to find rules of international humanitarian law that do not directly or indirectly protect rights of protected persons in international armed conflicts. In both international and non-international armed conflicts, those rules furthermore protect "basic rights of the human person" which are classic examples for jus cogens (Sassòli, 2002, p. 414).

In conclusion, if the breach concerns a basic rule of IHL or the prohibition of arbitrary deprivations of life, *force majeure* does not apply.

6.2 Assurance of Non-Repetition

According to article 30 of the ARSIWA, a State that committed a wrongful act is obliged to cease the act and offer guarantees and assurances of non-repetition if the circumstances so require (UN ILC 2001c, art. 30). As stated in ILC commentaries, this assurance is not always needed and aims to restore confidence and prevention (UN ILC, 2001b, pp. 89–91).[6] "Assurances are normally given verbally, while guarantees of non-repetition involve something more—for example, preventive measures to be taken by the responsible State designed to avoid repetition of the breach" (UN ILC, 2001b, p. 90).[7]

---

6 ILC comments that "(9) Subparagraph (b) of Article 30 deals with the obligation of the responsible State to offer appropriate assurances and guarantees of non-repetition, if circumstances so require. Assurances and guarantees are concerned with the restoration of confidence in a continuing relationship, although they involve much more flexibility than cessation and are not required in all cases. They are most commonly sought when the injured State has reason to believe that the mere restoration of the pre-existing situation does not protect it satisfactorily. (…) Such demands are not always expressed in terms of assurances or guarantees, but they share the characteristics of being future-looking and concerned with other potential breaches. They focus on prevention rather than reparation and they are included in article 30" (UN ILC, 2001b, pp. 89–90).
"(13) In some cases, the injured State may ask the responsible State to adopt specific measures or to act in a specified way in order to avoid repetition. Sometimes the injured State merely seeks assurances from the responsible State that, in future, it will respect the rights of the injured State. In other cases, the injured State requires specific instructions to be given, or other specific conduct to be taken. But assurances and guarantees of non-repetition will not always be appropriate, even if demanded. Much will depend on the circumstances of the case, including the nature of the obligation and of the breach. The rather exceptional character of the measures is indicated by the words "if circumstances so require" at the end of subparagraph (b). The obligation of the responsible State with respect to assurances and guarantees of non-repetition is formulated in flexible terms in order to prevent the kinds of abusive or excessive claims which characterized some demands for assurances and guarantees by States in the past" (UN ILC, 2001b, pp. 90–91).

7 Regarding the ICJ's decisions, we can observe that the Court seems to be reluctant to grant assurances of non-repetition (Barbier, 2010, pp. 554–555).

(Contd.)

The Basic Principles and Guidelines on the Right to a Remedy and Reparation for Victims of Gross Violations of IHRL and Serious Violations of IHL states that victims of serious violations of IHL should be provided with all-encompassing and adequate reparation, including guarantees of non-repetition, according to the specific circumstances of each case and proportionally to the breach. Within the guarantees of non-repetition, it cites "Ensuring effective civilian control of military and security forces" and "Ensuring that all civilian and military proceedings abide by international standards of due process, fairness and impartiality," "Promoting the observance of codes of conduct and ethical norms, in particular international standards, by public servants, including law enforcement, correctional, media, medical, psychological, social service, and military personnel, as well as by economic enterprises" among others (UNGA, 2005).[8]

We recall the La Grand Case (ICJ, 2001) in which the ICJ, considered that the commitment undertaken by the USA "to ensure implementation of the specific measures adopted in performance of its obligations under Article 36, paragraph 1 (b), of the Vienna Convention on Consular satisfied Germany's claim for a general assurance of non-repetition" (ICJ,2001). In the judgment that was issued in the same year as the ARSIWA, the ICJ made no reference to ILC's work.

In Cameroon v. Nigeria, the Court did not uphold Cameroon's claim for assurances of non-repetition.

"318. Cameroon, however, is not only asking the Court for an end to Nigeria's administrative and military presence in Cameroonian territory but also for guarantees of non-repetition in the future. Such submissions are undoubtedly admissible (LaGrand (Germany v. United States of America), Judgment, I.C.J. Reports 2001, pp. 508 et seq., paras. 117 et seq.). However, the Judgment delivered today specifies in definitive and mandatory terms the land and maritime boundary between the two States. With all uncertainty dispelled in this regard, the Court cannot envisage a situation where either Party, after withdrawing its military and police forces and administration from the other's territory, would fail to respect the territorial sovereignty of that Party. Hence Cameroon's submissions on this point cannot be upheld" (ICJ 2002, par 318).

In Legal Consequences of the Construction of a Wall in the Occupied Palestinian Territory Advisory Opinion, the Court noted it was contended that Israel shall offer assurances on non-repetition but did not address this question.

"145. As regards the legal consequences for Israel, it was contended that Israel has, first, a legal obligation to bring the illegal situation to an end by ceasing forthwith the construction of the wall in the Occupied Palestinian Territory, and to give appropriate assurances and guarantees of non-repetition" (ICJ, 2004, par 145).

In Djibouti v. France, the ICJ also noted that Djibouti requested assurances of non-repetition but did not elaborate or decide on the topic.

"11. That the French Republic shall provide the Republic of Djibouti with specific assurances and guarantees of non-repetition of the wrongful acts complained of" (ICJ, 2008).

In the jurisdictional Immunities of the State case, the ICJ stated that it must be presumed that States act in good faith and did not request Italy to ensure non-repetition "(...) while the Court may order the State responsible for an internationally wrongful act to offer assurances of non-repetition to the injured State, or to take specific measures to ensure that the wrongful act is not repeated, it may only do so when there are special circumstances which justify this, which the Court must assess on a case-by-case basis. In the present case, the Court has no reason to believe that such circumstances exist" (ICJ, 2012).

8 According to the document,

"18. In accordance with domestic law and international law, and taking account of individual circumstances, victims of gross violations of international human rights law and serious violations

(Contd.)

The question posed here is whether assurances or guarantees of non-repetition are possible with AWS, which may act unpredictably and self-learn. Here, issues of attribution previously discussed will be set aside once more.

A State could guarantee, for instance, not to use a specific AWS or algorithm. It could also provide venues for civil scrutiny of AWS and promote the observance of codes of conduct by public servants and private enterprises or enact directives as regards AWS. Ensuring that AWS would not repeat a breach outside those parameters might be challenging. Considering the inherent degree of unpredictability of AWS, as explained in Part I, it is almost impossible to ensure non-repetition. We envision some preventive measures, such as improving programming and testing. Nonetheless, such measures might be better called safeguards and preventions rather than guarantees and assurances. If a State were to settle an agreement or a court to rule on assurances of non-repetition, it would be prudent to clarify this caveat.

---

of international humanitarian law should, as appropriate and proportional to the gravity of the violation and the circumstances of each case, be provided with full and effective reparation, as laid out in principles 19 to 23, which include the following forms: restitution, compensation, rehabilitation, satisfaction and guarantees of non-repetition."

"23. Guarantees of non-repetition should include, where applicable, any or all of the following measures, which will also contribute to prevention:

(a) Ensuring effective civilian control of military and security forces;

(b) Ensuring that all civilian and military proceedings abide by international standards of due process, fairness and impartiality;

(c) Strengthening the independence of the judiciary;

(d) Protecting persons in the legal, medical and health-care professions, the media and other related professions, and human rights defenders;

(e) Providing, on a priority and continued basis, human rights and international humanitarian law education to all sectors of society and training for law enforcement officials as well as military and security forces;

(f) Promoting the observance of codes of conduct and ethical norms, in particular international standards, by public servants, including law enforcement, correctional, media, medical, psychological, social service and military personnel, as well as by economic enterprises;

(g) Promoting mechanisms for preventing and monitoring social conflicts and their resolution;

(h) Reviewing and reforming laws contributing to or allowing gross violations of international human rights law and serious violations of international humanitarian law" (UNGA, 2005, arts. 18 and 23).

CHAPTER

7

Autonomous Weapons Systems and Special Issues Related to Primary Norms Impacting Responsibility

This chapter focuses on AWS issues related to primary norms that impact State responsibility but that are not explicit in the ARSIWA, namely human-machine interaction (7.1), responsibility for not using AWS (7.2), AWS review and State responsibility (7.3), and State responsibility for breach of due diligence in the context of AWS (7.4). As in all chapters of Part II, the focus will be on existing norms (*de lege lata*), as (*de lege ferenda*) possibilities are mainly addressed in Part III. However, considering that we are dealing with primary norms that impact State responsibility on AWS-related issues and that there are no primary norms on AWS, it was necessary to include some prospective options to develop the analyses.

7.1 Human-Machine Interaction and State Responsibility

This section aims to provide an overview of the debate on human-machine interaction as a critical element of the emerging regulation of AWS and then discuss the repercussions of this emerging requirement on State responsibility. The guiding question is: How does the human-machine interaction impact State responsibility? First, we will discuss CCW GGE guiding principles and States' positions on the issue. Following, we will turn to the obligation to ensure a human-machine interaction and how human-machine interactions impact State responsibility.

7.1.1 CCW GGE's Principles Related to Human-Machine Interaction

In 2018, States at CCW GGE agreed by consensus on 10 Guiding Principles on AWS, and in 2019, the 11th Guiding Principle was agreed on (GGE CCW

2018; GGE CCW 2019a). Principles B, C, and K deal with the human-machine interface.

According to Guiding Principle B, "Human responsibility for decisions on the use of weapons systems must be retained since accountability cannot be transferred to machines. This should be considered across the entire life cycle of the weapons system" (GGE CCW 2019a). Therefore, at least in a broad sense, a human must decide on using AWS. However, the principle does not require a human to be responsible for every targeting decision taken in the context of AWS. The principle focuses on human responsibility and does not consider State responsibility.

Guiding Principle C states that accountability for developing, deploying, and using AWS "must be ensured in accordance with applicable international law, including through the operation of such systems within a responsible chain of human command and control" (GGE CCW 2019a). Like Guiding Principle B, it is also focused on human responsibility. However, a responsible chain of command and control, one of the methods Guiding Principle C suggests for ensuring accountability, might not be (at least directly) applicable to AWS. As raised by Bruun, "the question seems to be what does a responsible chain of command and control look like when you use AWS, and our question is also to States to what extent AWS reshuffles the decision-making structures States have in place" (Bruun, personal communication, June 21, 2022). Our view is that AWS change the formerly used decision-making paradigm, as devices can now perform tasks that previously only humans could do. Therefore, the suggested chain of command and control might be broken through the use of AWS. If this principle is to be used to ensure responsibility, States must clarify what is a responsible chain of command and control for the purposes of AWS.

Guiding Principle K recognizes that human-machine interaction might take diverse forms and be implemented throughout the AWS lifecycle. This interaction must comply with international law and IHL. "In determining the quality and extent of human-machine interaction, a range of factors should be considered, including the operational context, and the characteristics and capabilities of the weapon system as a whole" (GGE CCW 2019a). It suggests that human-machine interaction is needed by its mode, and the extent must be defined on a case-by-case basis.

In summary, none of the principles above expressly mention State responsibility. All three principles point to a need for some human involvement in the process of using AWS. Furthermore, as principles, they provide general guidelines but not a clear path on how to implement them.

7.1.2 Meaningful Human Control Discussions at CCW GGE

Meaningful Human Control (MHC) is a broad framework concept that discusses the human role in human-AWS interaction. Its specific content is yet to be construed. MHC is in the spotlight of both scholarship and States as a possible venue to address some AWS challenges, including responsibility. The focus on the human element enhances the debate over individual responsibility and might, paradoxically, have two side effects: contribute to shielding State responsibility or advancing State responsibility.

The ICRC emphasized at CCW GGE that the need to maintain human responsibility or control over AWS is a common ground on which States have been building agreements (ICRC, 2019c). In the same sense, the European Parliament issued a recommendation calling for the urgent negotiation of an international ban on weapon systems that lack human control over the use of force (European Parliament, 2018). The EU's position is that "Humans must make the decisions with regard to the use of lethal force, exert control over the lethal weapons systems they use, and remain accountable for decisions over life and death" (European Union External Action Service, 2018).

While the UK, France, China, and Korea supported MHC at CCW GGE, the US, Russia, and Israel argued for a lower threshold of human-machine involvement.

The UK stated that GGE's focus on "human control appears to be the most productive way forward, given the continuing lack of consensus on definitions, characteristics and even whether LAWS exist" (UK delegation, 2018). MHC enables military effectiveness and aids in preventing unintended, undesirable outcomes. However, MHC is a complex concept that must be considered throughout the AWS lifecycle. GGE on LAWS should aim at agreeing on "what elements of control over weapon systems should be retained by humans" (UK delegation, 2018). The UK suggests that MHC is discussed jointly with a compilation of good practices (UK delegation, 2020). In 2021, the UK stressed the need for discussion on MHC, "regarding how it manifests, what makes it meaningful, and the forms by which humans interact with AWS. It argued that human control is complex, situation-dependent, dynamic, and not bound to a single moment in time" (Acheson, 2021). In 2022, while proposing a manual on AWS, the UK stated that Human-Machine Interaction "would be an overarching theme throughout the whole document. The document would need to address the level of human involvement necessary to achieve the IHL ends" (UK delegation, 2022).

France argues that to comply with the eleven Principles, especially Principles B, C, D and G, States must ensure "that humans will remain responsible for designing, programming, defining and validating the rules of engagement, the rules for use and the operating rules of weapons systems based on emerging technologies in the area of LAWS" (French delegation, 2020). They also stated that LAWS that are fully autonomous should not be developed, this means weapons systems that can act "without any form of human supervision or dependence on a command chain by setting their own objectives or by modifying, without any human validation, their initial programme (rules of operation, use, engagement) or their mission framework" (French delegation, 2020).

South Korea claimed that MHC and human-machine interaction must be implemented throughout the diverse phases of any AWS lifecycle and that this interaction might develop depending on technological advances (Acheson, 2021).

China recognized how general the concept of MHC is and called for further elaboration and clarification. "Discussions on Human-Machine Interaction should first have a clear definition on LAWS and secondly define the mode and degree of human involvement and intervention" (China's Delegation, 2018). They claimed that "necessary measures of Human-Machine Interaction are conducive to the prevention of indiscriminate killing and maiming by LAWS caused by a breakaway from human control" (China's Delegation, 2018). During the 2021 August meeting, "China said states must decide how and to what extent humans should be involved in the development and use of WS to prevent them from being indiscriminate" (Acheson, 2021).

The 2022 Proposal of Draft Protocol IV submitted by Argentina, Ecuador, Costa Rica, Nigeria, Panama, the Philippines, Sierra Leone, and Uruguay also uses the threshold of MHC (Argentina and others, 2022b).[1]

The US, Russia, and Israel rejected the concept of MHC in the 2021 August meeting (Acheson, 2021, p. 13).[2]

The US does not argue for MHC: "The concept of "human control" is subject to divergent interpretations that can hinder meaningful discussion. 47. (…) we believe that emphasis on "control" would obscure rather than clarify the genuine challenges in this area" (US Delegation, 2018a). The US claims for "appropriate levels of human judgment," which is a lower threshold concept and "reflects a deliberate decision to permit weapons that are programmed to make "decisions" that relate to targeting" (US Delegation, 2018a, p. 2).

> "'Appropriate' is a flexible term that reflects the fact that there is not a fixed, one size-fits-all level of human judgment that should be applied to every context. What is 'appropriate' can differ across weapon systems, domains of warfare, types of warfare, operational contexts, and even across different functions in a weapon system. Some functions might be better performed by a computer than a human being, while other functions should be performed by humans" (US Delegation, 2018a, p. 2).

Israel used the term human judgment and not human control and affirmed that concerning the "(...) human involvement needed to ensure compliance with IHL, Israel would initially state that human judgment will always be an integral

---

1 According to their submission, it "refers to the threshold of application of human judgment and intervention necessary to ensure the maintenance of human agency, responsibility, proportionality and accountability in undertaking decisions regarding the use of any weapon and the ability of human operators to effectively supervise any weapon, undertake the necessary interaction that could either be directive or preventive, and to deactivate, terminate, or abort the operation of the weapon altogether" (Argentina and others, 2022b).

2 Acheson observes that "On the other hand, Israel, Russia, and the United States (USA) rejected the concept of HC. USA prefers the term human-machine interaction (HMI), which it said needs to be examined across the WS lifecycle. Israel, meanwhile, said human judgment is an integral part of LAWS across their lifecycles" (Acheson, 2021, p. 13).

part of any process regarding LAWS, and will be applied throughout their life-cycle" (Israel Permanent Mission to the UN Geneva, 2019b).

Russia's position is ambiguous. In November 2019, it argued that the concepts of "human control" and "human involvement" require subjective assessments and are irrelevant (Wareham, 2020). Nonetheless, in 2020, Russia reaffirmed "(…) its commitment to the need to maintain human control over the so-called LAWS, no matter how advanced these systems may be. It is human responsibility to ensure compliance with IHL norms during the combat use of the so-called LAWS" (Russian Federation Delegation, 2020). They highlighted that MHC is essential to implement Guiding Principles B and D and stated that according to their domestic law, "(…) a responsible official is always accountable for decisions concerning development and use of weapons, including emerging technologies in the area of LAWS" (Russian Federation Delegation, 2020). In the 2021 meeting, Russia rejected the concept of human control. It affirmed that "MHC is a problematic concept without clear definition, parameters, or agreed criteria. It said the form and method of human control should be left to states to decide but also argued that determining the degree of human control would be fully subjective in nature and be biased toward narrow national interests" (Acheson, 2021).

In our view, despite some actors treating it as a synonym, there is a difference between control, which requires the capability to abort the action, activate and deactivate the system (Moyes, 2013),[3] and judgment, as judgment requires less from the human, meaning that humans shall act to the best of their judgment (Garcia, personal communication, June 13, 2022).[4] The focus on MHC is relevant to ensuring individual responsibility and might be the best alternative both from an ethical and juridical standpoint. Nonetheless, the technological reality seems to navigate in another direction. AI devices are performing tasks humans used to do, such as driving cars and firing at targets. There might be an actual situation in

3 Moyes states that MHC requires:
"• Information – a human operator, and others responsible for attack planning, need to have adequate contextual information on the target area of an attack, information on why any specific object has been suggested as a target for attack, information on mission objectives, and information on the immediate and longer-term weapon effects that will be created from an attack in that context. • Action – initiating the attack should require a positive action by a human operator. • Accountability – those responsible for assessing the information and executing the attack need to be accountable for the outcomes of the attack" (Moyes, 2013).

4 This is a citation from the author's Ph.D. dissertation. During an interview for the author's Ph.D., Eugênio Vargas Garcia, in his personal capacity, stated that: "Meaningful human control is different from judgment, it has a subtlety of terms. The United States used a lot of human judgment. Judgment is weaker than human control. Controlling you intervene and can not only activate but deactivate the system. You must be able to abort or stop an action when it becomes dangerous, before you commit a violation. Judgment is much more malleable." My translation from the original in Portuguese:
"Controle humano significativo é diferente de julgamento, tem uma sutileza dos temos. Os Estados Unidos usavam bastante julgamento humano. Julgamento é mais débil que controle humano. Controlar você intervém e pode não só ativar como desativar o sistema. Você deve ser capaz de abortar ou interromper uma ação quando ela se torna perigosa, antes que você cometa uma violação. Julgamento é muito mais maleável" (Garcia, personal communication, June 13, 2022).

which no human might be able to exercise meaningful control. In those situations, requiring MHC might equal a ban on AWS. In this scenario, perspectives such as that of the US that claim appropriate levels of human judgment and focus on a meta-decision, which means the decision to allow the autonomous device to decide, seem more prone to be accepted and implemented by the international community, at least as a first step. From a similar perspective, Cummings states that in this technological, speedy, and complex warfare framework, war fighters often have to decide between life and death without the necessary skills and appropriate time for reflection, which leads to a high chance of mistakes.[5] She concludes, "The key to this future is meaningful human certification of autonomous weapons, not insisting on an illusory concept of meaningful human control" (Cummings, 2019, p. 20).

In short, States such as China, France, Korea, and the UK claim for MHC, at least in the discourse. In contrast, Israel and the US call for human judgment. All the States analyzed argue that some human-machine interaction is required. Our view is that despite being ethically and juridically desirable to require MHC, it might not be feasible. Requiring appropriate levels of human judgment as a minimum standard is implementable and encounters no opposition from States. Appropriate levels of human judgment are also aligned with the three Guiding Principles discussed in the previous section.

7.1.3 Meaningful Human Control, Appropriate Levels of Human Judgment, and State Responsibility: An Analysis of Prospective Paths

When a breach occurs, both individual responsibility and State responsibility might arise. They are, therefore, two sides of the same coin. Human-machine interaction impacts both and will be analyzed with a focus on State responsibility (Bruun, personal communication, June 21, 2022).[6]

---

5 Cummings stated that "Proponents of a ban on offensive autonomous weapons advocate that any use of such weapons should not be beyond meaningful human control (Future of Life Institute, 2015). This language is problematic at best because there are widely varying interpretations of what meaning human control is. For example, meaningful human control (MHC) could mean that a human has to initiate the launch of such a weapon. In today's networked world, this person can (and often is) thousands of miles from the intended target. It hardly seems meaningful for a human to release a weapon if he or she is sitting 4000 miles from the point of weapon release. MHC might mean a human has to monitor a weapon until impact and perhaps have the ability, however remotely, to abort the mission. In such cases, as will be discussed in more detail, we are often asking humans to take on MHC tasks that have high likelihoods of human error with deadly unintended consequences" (Cummings, 2019, p. 20).

6 This is a citation from the author's Ph.D. dissertation. During an interview for the author's Ph.D., Laura Bruun, in her personal capacity, stated: "The linkage has to be made between State responsibility and human control because discussions around human control are very relevant if you want to inform discussion around how to ensure State responsibility. If you don't know what rules of IHL require from humans you don't know if the State committed a breach. So, it is two sides of the same coin, but I think the link has not really been addressed. States did not approach the topic with state responsibility in mind. That link could definitely be strengthened" (Bruun, personal communication, June 21, 2022).

First, we will address the question of the impacts of the human-machine relation on State responsibility and discuss MHC, appropriate levels of human judgment, and State responsibility. Next, we will turn to whether there is an international obligation for States to ensure a degree of human involvement and State responsibility arising thereof.

From the perspective of individual responsibility, suppose there is a requirement of MHC, and a violation occurs. The person(s) in charge of the design or that authorised the design could be held responsible. MHC ensures that there will be a responsible person, thus leading to strict State responsibility. In this scenario where individual responsibility attaches to a State's agent, attributing responsibility to the State is not challenging. However, States such as the US, Russia, and Israel oppose it and claim a human-machine relation along the lines of appropriate levels of human judgment.

On the other hand, if there is an appropriate level of human judgment and a breach occurs, and the judging person states that they acted according to the best of their knowledge, the individual will not be held responsible. Therefore, the human agent's responsibility is subjective, and only those who did not act to the best of their knowledge can be held responsible. In that case, State responsibility is clear-cut if the individual did not act according to the best of their judgment. A further question is whether appropriate levels of human judgment entail strict State responsibility in this hypothesis. In other words, can an event be deemed an accident for the human agent and an internationally wrongful act for the State? In our view, the general rule is that State responsibility is strict if one of its agents, despite acting according to the best of their knowledge regarding the appropriate level of human judgment, generated the breach. This is the general rule that follows from our proposed interpretation of the ARSIWA, which does not require intent, and "In the absence of any specific requirement of a mental element in terms of the primary obligation, it is only the act of a State that matters, independently of any intention" (UN ILC, 2001b, p. 36). Furthermore, many domestic legal systems entail objective responsibility of the State for their agents' misdoings.[7] However, if the primary rule requires intent, the human-machine interaction impacts whether or not a breach of an international obligation has occurred.

> Primary rules of IHL are unclear in terms of how do you define an inherently indiscriminate weapon, for instance, and what types of foresee ability and what type of in general human control do you need. This is why the discussion on human control is also important to identify whether a State committed a breach (Bruun, personal communication, June 21, 2022).

Suppose State responsibility requires intent due to the content of the primary rule. In that case, we claim for a lower threshold, meaning that recklessness or negligence suffices for State responsibility to ensue, which is based on the State's

7 Which is the case in Brazil, for example.

due diligence obligation that will be further discussed in the next section and on the inherent risk of activities involving AWS. When there is a requirement for human judgment and State agents authorize the attack, even without authorizing every individual decision, States are responsible for the acts of their AWS.

Considering that the proposed responsibility framework depends on the future adoption of a framework of MHC or appropriate levels of human judgment and could be interpreted differently, a statement *de lege ferenda* is necessary.

7.1.4 Is there an International Obligation for States to Ensure a Degree of Human Involvement?

Turning to the initial question of whether there is an international obligation for States to ensure a degree of human involvement regarding AWS, the answer is yes, based on the AWS guiding principles, States' positions, and IHRL and IHL.[8] We add that UNESCO's Recommendation on the Ethics of AI, which focuses on non-military issues, refers to the necessity of human supervision, embracing individual and public oversight, and requires that decisions to rely on AI systems in limited conditions must be human (UNESCO, 2021).[9] Also corroborating the necessity of human involvement in the process is the Protocol amending the Convention for the Protection of Individuals concerning the Automatic Processing of Personal Data. It foresees the right "not to be subject to a decision significantly affecting him or her based solely on an automated processing" and "to obtain, on request, knowledge of the reasoning underlying data processing where the results of such processing are applied to him or her" (Protocol amending the Convention for the Protection of Individuals with regard to Automatic Processing of Personal Data, 2018, art. 11).

---

8 The grounding principle of IHRL is human dignity, and the grounding principle of IHL is humanity. Therefore, despite different historical origins, both are concerned with protecting the human person (Mauri, 2022, p. 67). On IHRL, for instance, article 1 of the UDHR: "All human beings are born free and equal in dignity and rights. They are endowed with reason and conscience and should act towards one another in a spirit of brotherhood." Also, IHRL right to life entails the prohibition of arbitrary deprivation of life "For the killing of a human to be meaningful, it must be intentional. That is, it must be done for reason and purpose. (…) If a combatant is to die with dignity, there must be some sense in which that death is meaningful. In the absence of an intentional and meaningful decision to use violence, the resulting deaths are meaningless and arbitrary, and the dignity of those killed is significantly diminished" (Asaro, 2016, p. 385). On IHL among the relevant principles are the principle of humanity, which is expressed in the Martens Clause, and the principles of Distinction and Proportionality.

9 UNESCO's Recommendation on the Ethics of Artificial Intelligence state that: "35. Member States should ensure that it is always possible to attribute ethical and legal responsibility for any stage of the life cycle of AI systems, as well as in cases of remedy related to AI systems, to physical persons or to existing legal entities. Human oversight refers thus not only to individual human oversight, but to inclusive public oversight, as appropriate.
36. It may be the case that sometimes humans would choose to rely on AI systems for reasons of efficacy, but the decision to cede control in limited contexts remains that of humans, as humans can resort to AI systems in decision-making and acting, but an AI system can never replace ultimate human responsibility and accountability. As a rule, life and death decisions should not be ceded to AI systems" (UNESCO, 2021).

There seems to be an international consensus that some level of human involvement is necessary, but there is little agreement about the degree of human involvement needed. This means that zooming out, the decision ought, at some point, to be made by a human. However, zooming in does not necessarily mean that every individual targeting decision needs to be made by a person, as no weapon's prohibition conditions its legality on the possibility of blaming an individual (Dunlap, 2016, p. 66).

A minimum core around a lower threshold of appropriate levels of human judgment seems to exist, which finds no opposition in the international arena. This minimum standard also resonates with the guiding principles and IHRL and IHL norms. Whereas the MHC paradigm seems more ethical and more protective of human dignity, it might not be feasible, finds opposition in the international community, and is not an option *de lege lata.*

Based on the international consensus that, at a minimum, States have to ensure appropriate levels of human judgment, if States fail to do so, they can be held responsible for not complying with this obligation.

In summary, under the minimum threshold of appropriate levels of human judgment, we claim that States should respond under strict liability for the breaches, whereas for responsibility to attach to State agents requires *mens rea.* If the primary rule requires intent, States shall be held responsible if they act negligently or recklessly. Furthermore, considering the claimed obligation to ensure appropriate levels of human judgment, States might be held responsible if they fail to do so. Considering that the proposed responsibility framework could be interpreted differently, a statement *de lege ferenda* is necessary.

A challenge is to define what appropriate levels of human judgment mean and how to test if that level has been sufficiently met. *De lege ferenda*, the development of ontological patterns[10] is an alternative to provide certainty on what appropriate levels of human judgment mean and require, as will be discussed in Part III (Prestes, personal communication, July 14, 2022).[11]

---

10 Ontological standards are models that represent, by formally naming, and defining concepts, relations and properties, and the relation among them. See for example (IEEE, 2021a).

11 This is a citation from the author's Ph.D. dissertation. During an interview for the author's Ph.D., Edson Prestes, in his personal capacity, talking about MHC and not about appropriate levels of human judgment, stated that: "That's where ontological patterns come in. They are important for giving more precision to real-world concepts and definitions. When we have a definition expressed in natural language, that definition, [as in the case of meaningful human control] is very subject to ambiguity, and different interpretations. For example, what is significant in human control? Can we measure it? If so, at how many levels?" My translation from the original in Portuguese:
"É aí que entram os padrões ontológicos. Eles são importantes para dar mais precisão a conceitos e definições do mundo real. Quando a gente tem uma definição expressa em linguagem natural, essa definição, [como no caso de controle humano significativo] é muito sujeito a ambiguidade, e diferentes interpretações. Por exemplo, o que é significativo no controle humano ? Podemos mensurá-lo? Se sim, em quantos níveis?" (Prestes, personal communication, July 14, 2022).

7.2 Responsibility for not Using AWS

In this section, the question is if, within the framework of the ARSIWA, a State can be held responsible for not using AWS if it is proven that such use could reduce causalities and save lives.

We will briefly present the consequentialist[12] perspective to better comprehend responsibility for not using AWS.

7.2.1 The Consequentialist Perspective

Consequentialists determine the morality of decisions using the states of affairs the choices produce as a benchmark (Alexander and Moore, 2021). Consequentialists maintain that AWS should not be banned if they can produce better results than humans on the battlefield. For utilitarians, humanitarian considerations require stimulating the development and use of AWS.[13]

From an ethical standpoint, consequentialists believe AWS will outperform humans since, for instance, they do not consider self-preservation a leading factor, do not have emotions that could blur the decision-making process, and are capable of using more sensors and processing information much faster than humans (Arkin, 2009, p. 30). Moreover, they do not have the human bias of 'scenario fulfillment,' which is the "distortion or neglect of contradictory information in stressful situations, where humans use new incoming information in ways that only fit their pre-existing belief patterns, a form of premature cognitive closure" (Arkin, 2009, p. 30). Thus, from a consequentialist standpoint, AWS should be used if they can reduce violations.

In a similar vein, but from a juridical standpoint, Sassòli argues that considering that the lives of people who deploy AWS are not at risk, their use might create possibilities for additional protection. For instance, AWS are not motivated to save themselves and can 'shoot second.'[14] Sassòli also recalls that AWS, with machine learning, learn from experience how to improve

---

[12] The moral philosophy grounded debate between consequentialist and deontological perspectives has taken place mainly among those who discuss whether AWS must be banned or not. On one extreme of the spectrum is the deontological approach that envisions that even if AWS can behave better than humans and its outputs conform to IHL, they should be banned as a matter of legal or ethical principles. In this sense: (Lieblich and Benvenisti, 2016, p. 246). Heyns claims that "(...) there are limits beyond which our transfer of power to technology should not go – not on the battlefield, and not in other realms of life. The weapon cannot be allowed to become the warrior" (Heyns, 2017, p. 60).
On the other extreme of the spectrum is the consequentialist's perspective, which will be considered in section 8.2.1.

[13] Lieblich and Benvenisti stated that "If we all take humanitarian considerations seriously, we should encourage the development and use of such weapons" (Lieblich and Benvenisti, 2016, p. 246).

[14] Nonetheless, he recognizes this does not prevail if the opposite party uses anti-AWS devices, in which case AWS will likely be programmed to shoot first to prevent its destruction (Sassòli, 2014, pp. 335–336).

precautionary measures (Sassòli, 2014, pp. 335–336).[15] Imagine a conflict in which the two parties use only AWS. Wouldn't such a conflict be preferable to a conflict between humans? It would be in line with the obligation to minimize lethality and suffering. Moving further and highlighting the precaution concern, Margulies argues that "states should use AI-based situation awareness technology (SAT) as a feasible precaution that will make target selection more accurate and reduce harm to civilians" (Margulies, 2018). Michael Schmitt also envisions this obligation if AWS increases compliance with IHL, as humans might not be as precise as the multiple sensors of AWS to deal with the armed conflict scenario (Schmitt, 2015).[16] Furthermore, he expresses a legal requirement to use available AWS if they prevent or minimize harm. "Those who would ban the system must understand that commanders will consequently be deprived of a means of warfare that might avoid civilian casualties, one that they would otherwise have to employ as a matter of law" (Schmitt, 2015).

In short, the argument posed by consequentialists is that AWS can add precaution to an attack since they may outperform humans in preventing causalities (Pourcel, 2018, pp. 81–82; Anderson and Waxman, 2013, p. 17),[17] thus

---

15 Marco Sassòli stated that "Intelligent weapons have an additional advantage. The feasibility of precautions evolves through experience. When precautions taken in the past proved unsuccessful, that may imply the need to learn lessons (and belligerents have, in my view, an obligation to foresee pertinent procedures) to avoid such incidents in the future. If artificial intelligence can be created, it is essential to make sure that weapons with such intelligence can be recalled and reprogrammed, and that human beings monitor the development of that intelligence in order to quickly take advantage of lessons learned" (Sassòli, 2014, pp. 335–336).

16 In a similar vein, during an interview for the author's Ph.D., Diego Mauri, in his personal capacity, when asked about State responsibility for not using AWS, stated that: "You have a general duty of precaution according to IHL. So, you can push yourself to argue that if you dispose of this technology, there's a kind of an obligation. I mean, if it's proved that if you use that AWS would lower the margin of errors then the responsibility can follow. It's an argument that can be supported. This is what I think I'm not that straightforward. I wouldn't say yes to 100% or no to 100% percent. I think it's a matter of interpretation" (Mauri, personal communication, July 5, 2022). This is a citation from the author's Ph.D. dissertation.

17 Pourcel states that "(...) it is quite possible that the designers of DRIA will be able to demonstrate or contrary to what is affirmed that an autonomous weapon system "stocked" with the rules of international humanitarian law and the law of war, the rules of engagement additional particulars specific to each compliant country or the law of armed conflict, image data illustrating its application and data imposing compliance with the protocol like the OODA loop will undoubtedly be technically capable of avoiding the collateral damages." My translation from the original in French: "(...)il est fort possible que les concepteurs de DRIA soient capables de démontrer ou contraire de ce quiest effirmé quún systèmearmé autonome "achalandé" des règles du droit international humanitaire et du droit de la guerre, des règles déngagement partoculières supplétives propes à chaque pays conformes ou droit des conflits armés, de données images illustrant son application et de données imposant le respect de protocole à l'instar de la boucle de OODA sera sans doute capable techniquement d'éviter mieux que l'homme les dommages collatéraux (...)" (Pourcel, 2018, pp. 81–82).

Anderson and Waxman, state that "Excessive devotion to individual criminal liability as the presumptive mechanism of accountability risks blocking development of machine systems that might, if successful, reduce actual harms to soldiers as well as to civilians on or near the battlefield. Effective adherence to the law of armed conflict traditionally has been through mechanisms of state (or armed party) responsibility. Responsibility on the front end, by a party to

(*Contd.*)

reducing harm to soldiers and civilians. Within the consequentialist framework presented, State responsibility for not using AWS can be discussed.

7.2.2 State Statements at CCW that Lean towards a Consequentialist Perspective

The Russian Federation, the US, and Israel presented their views at the CCW GGE meeting, which lean towards a consequentialist perspective.

Russia, commenting on GGE Guiding Principle H, which states that "Considerations should be given to the use of emerging technologies in the area of lethal autonomous weapons systems in upholding compliance with IHL and other applicable international law obligations," confirmed that:

> (…) the so-called LAWS can significantly reduce the negative effects of the use of weapons in the context of IHL. The existing complexes of a high degree of military autonomy in the Russian Federation significantly contribute to the compliance during hostilities with such key principles of IHL as proportionality and distinction. This is due to the fact that, in addition to their technological advantages (accuracy, speed, effectiveness), such weapons neutralize human-caused risks (operator's mistakes due to his or her mental or physiological state, ethical, religious or moral attitudes), and thus reduce the probability of unintentional attacks against civilians and non-military targets (Russian Federation Delegation, 2020).

Russia underlined that AWS could potentially be more effective than human agents as AWS can reduce the chances of failures."In particular, such systems are capable of considerably reducing the negative consequences of the use of weapons related to operator's errors, mental and physiological state, as well as ethical, religious or moral stance in the IHL context" (Slijper, Beck, and Kayser, 2019). In the same sense, the Russian 2022 paper stressed that LAWS are devoid of weaknesses inherent in human beings (Russian Federation Delegation, 2022).

The US argues that the new "autonomy-related technologies," including AI and machine learning, such as self-driving cars and autonomous medical

---

a conflict, is reflected in how a party plans its operations, through its rules of engagement and the "operational law of war." Although administrative and judicial mechanisms aimed at individuals play some important enforcement role, the law of armed conflict has its greatest effect and offers the greatest protections in war when it applies to a side as a whole.

It would be unfortunate to sacrifice real-world gains consisting of reduced battlefield harm through machine systems (assuming there are any such gains) simply in order to satisfy an a priori principle that there always be a human to hold accountable. It would be better to adapt mechanisms of collective responsibility borne by a "side" in war, through its operational planning and law, including legal reviews of weapon systems and justification of their use in particular operational conditions" (Anderson and Waxman, 2013, p. 17).

Note that scholars such as Amoroso disagree with this perspective: "(…) due to the fact that the complete ouster of the human element from the targeting process would rule out a feasible precaution capable of preventing infringements of the principles of distinction and proportionality, as well as of the right to life under IHRL" (Amoroso, 2020, p. 107).

devices, are prone to improve or even save lives. In the same vein, autonomous technologies have the capability of saving lives in the context of armed conflicts. "This is especially the case because "smart" weapons that use computers and autonomous functions to deploy force more precisely and efficiently have been shown to reduce risks of harm to civilians and civilian objects" (US Delegation, 2018b). The US claims that discussions on the challenges AWS pose "(...) must involve consideration of how these technologies can be used to enhance the protection of the civilian population against the effects of hostilities" (US Delegation, 2018b). According to the US, State practice demonstrates that the use of AWS might reduce the risk to civilians, for example, by:

> "(1) incorporating autonomous self-destruct, self-deactivation, or self neutralization mechanisms; (2) increasing awareness of civilians and civilian objects on the battlefield; (3) improving assessments of the likely effects of military operations; (4) automating target identification, tracking, selection, and engagement; and (5) reducing the need for immediate fires in self-defense" (US Delegation, 2018b).

In sum, the US at CCW GGE expressed its belief that "(…) advances in autonomy and machine learning can facilitate and enhance the implementation of IHL, including the principles of distinction and proportionality." They present as the States' goal a more profound comprehension on "(…) how this technology can continue to be used to reduce the risk to civilians and friendly forces in armed conflict" (US Delegation, 2018c).

The US provided guiding questions to develop and deploy AWS, which shed light on the responsibility for not using AWS:

> "(a) Does military necessity justify developing or using this new technology? (b) Under the principle of humanity, does the use of this new technology reduce unnecessary suffering? (c) Are there ways this new technology can enhance the ability to distinguish between civilians and combatants? (d) Under the principle of proportionality, has sufficient care been taken to avoid creating unreasonable or excessive incidental effects? (e) Under the principle of the honor, does the use of this technology respect and avoid undermining the existing law of war rules?" (US Delegation, 2018a).

Israel stated that AWS might offer military and humanitarian advantages since they "(...) may serve to advance adherence to existing IHL, as technology may allow improving compliance with IHL. These may include better precision of targeting which would minimize collateral damage and reduce risk to combatants and noncombatants" (Israel Permanent Mission to the UN Geneva, 2020).

The US, Israel, and Russian arguments lean towards a utilitarian perspective, focusing on results and determining the proper conduct by evaluating which brings the greatest good for the greatest numbers.

7.2.3 Precision in the Attack: An Obligation to Use the More Precise Weapon?

Outside the context of AWS, humanitarian and disarmament groups have criticized States for using unguided munitions in populated areas, meaning they should use precision-guided munitions, a more advanced technology, when possible.[18] However, some of those same groups have advocated for a ban on AWS. First, we will briefly discuss the need to use more precise weapons. Next, we will discuss if AWS can be deemed the most precise weapon.

States have under IHRL and IHL the duty of prevention and precaution, and States must, as far as possible, minimize civilian harm, which impacts State options for choices of a weapon. States parties to IHRL treaties must respect, ensure respect, and protect the human rights of all human beings in their territory or falling under their jurisdiction (UNGA Human Rights Council, 2015; ICCPR, 1976, art. 2; UN ICCPRC, 2004; UN CRC, 2003). The Human Rights Council "Affirms the importance of effective preventive measures as a part of overall strategies for the promotion and protection of all human rights" (UNGA Human Rights Council, 2013). In the same sense, the IACHR reinforced that States have "a legal duty to take reasonable steps to prevent human rights violations" (IACHR, 1989, p. 31).

As stated at the Human Rights Council by the Joint report of the Special Rapporteur on the rights to freedom of peaceful assembly and of association and the Special Rapporteur on extrajudicial, summary, or arbitrary executions:

> 55. States are required to procure less lethal weapons for use in appropriate situations, with a view to increasingly restraining the application of means capable of causing death or injury. Less-lethal weapons must be subject to independent scientific testing and approval, and used responsibly by well-trained law enforcement officials, as such weapons may have lethal or injurious effects if not used correctly or in compliance with international law and human rights standards. States should work to establish and implement international protocols for the training on and use of less-lethal weapons (UNGA Human Rights Council, 2016).

Article 57 of 1977 Additional Protocol I (Additional Protocol I, 1977) states the obligation to "2(ii) take all feasible precautions in the choice of means and methods of attack with a view to avoiding, and in any event to minimizing, incidental loss of civilian life, injury to civilians and damage to civilian objects." In the same sense is rule 17 of the ICRC's Codification of Customary IHL: "Each party to the conflict must take all feasible precautions in the choice of means and methods of warfare with a view to avoiding, and in any event to minimizing, incidental loss of civilian life, injury to civilians and damage to civilian objects." (ICRC, n.d.b) Article 22 of the ICRC's Codification of Customary IHL also states

[18] See for instance (Docherty, 2022).

that "The parties to the conflict must take all feasible precautions to protect the civilian population and civilian objects under their control against the effects of attacks" (ICRC, n.d.c). Other IHL instruments, State military codes, and case law corroborate the need to use a more precise weapon when possible.[19]

[19] Examples of domestic and regional restraints on the choice of the means of warfare can be found at ICRC IHL Database Customary International Law, Practice Relating to Rule 17 (ICRC, n.d.b). Among the examples are (Second Protocol to the Hague Convention of 1954 for the Protection of Cultural Property in the Event of Armed Conflict, 1999, art. 7).
"France's LOAC Manual (2001) provides that those who plan or decide upon an attack shall take all precautions which are practically possible in the choice of means and methods of attack with a view to avoiding, and in any event to minimizing, loss of civilian life" (ICRC, n.d.b).
"Israel's Manual on the Rules of Warfare (2006) states:
The rules of war have laid down a number of rules of engagement in a theatre of war containing civilians:
- The means of attack should be planned in such a way as to prevent or at least minimize casualties among the civilian population
The Russian Federation's Regulations on the Application of IHL (2001) states:
Whenever there is a choice between several means and methods of engaging the enemy by fire, or between types of fire, for getting equal results preference is given to those means and methods … that pose the least danger to the civilian population and objects" (ICRC, n.d.b).
The UK LOAC Manual (2004) states: "There is the obligation to select the means (that is, weapons) or methods of attack (that is, tactics) which will cause the least incidental damage commensurate with military success."
The manual further states:
"With respect to attacks, the following precautions shall be taken:
(1) those who plan or decide upon an attack shall:
…
(b) take all feasible precautions in the choice of means and methods of attack with a view to avoiding, and in any event to minimizing, incidental loss of civilian life, injury to civilians and damage to civilian objects."
The US Naval Handbook (2007) states:
"[T]he commander must decide, in light of all the facts known or reasonably available to him, including the need to conserve resources and complete the mission successfully, whether to adopt an alternative method of attack, if reasonably available, to reduce civilian casualties and damage" (ICRC, n.d.b).
The International Criminal Tribunal for the former Yugoslavia:
In its judgment in the Kupreškić case in 2000, the ICTY Trial Chamber stated that Article 57 of the 1977 Additional Protocol I was now part of customary international law, not only because it specified and fleshed out general pre-existing norms, but also because it did not appear to be contested by any State, including those who had not ratified the Protocol.
With reference to the Martens Clause, the Trial Chamber held:
"The prescriptions of … [Article 57 of the 1977 Additional Protocol I] (and of the corresponding customary rules) must be interpreted so as to construe as narrowly as possible the discretionary power to attack belligerents and, by the same token, so as to expand the protection accorded to civilians."
The International Criminal Tribunal for the former Yugoslavia:
In its final report to the ICTY Prosecutor in 2000, the Committee Established to Review the NATO Bombing Campaign Against the Federal Republic of Yugoslavia stated:
"The military worth of the target would need to be considered in relation to the circumstances prevailing at the time. If there is a choice of weapons or methods of attack available, a commander should select those which are most likely to avoid, or at least minimize, incidental damage. In doing so, however, he is entitled to take account of factors such as stocks of different weapons and likely future demands, the timeliness of attack and risks to his own forces. In determining

(Contd.)

In summary, under IHL and IHRL there is a duty to use more precise weapons when possible.

7.2.4 Back to the Question: Responsibility for not Using AWS

Considering the panorama presented, we turn back to the question posed in this section: if, from a juridical standpoint, based on the ARSIWA, a State can be held responsible for not using AWS.

Cass R. Sunstein and Adrian Vermeule state that governments might be morally responsible for failing to prevent causalities by using AWS, depending on the circumstances.

> "If it could be shown that the use of FAWs would avert a certain number of unnecessary civilian casualties, it would seem problematic to argue that a rule mandating human involvement is paramount. Depending on where one comes down on the act/omission distinction, governments may even be morally responsible for failing to limit civilian casualties through the responsible use of FAWs" (Sunstein and Vermeule 2006, p. 747).

The authors touch upon the central question of this chapter regarding responsibility for not using AWS and answer it in the affirmative, but from a moral and not from a legal standpoint.[20] From a legal standpoint, supposing it is proven that AWS would be the more precise weapon, our view is that a State could be held responsible for not using AWS under the current international law.

If a State has at its disposal an AWS that has proven to have a high probability of preventing breaches of international law, a court might hold the State responsible for failing to use that AWS. Such a perspective would be based on the above consequentialist arguments and primary rules on precaution and prevention.

In summary, from a legal standpoint, *de lege lata*, considering IHL and IHRL obligations of precaution and prevention, and the obligation of using a more precise weapon, than not using AWS, if the State had at its disposal such AWS,

---

whether or not the mens rea requirement [intention or recklessness, for the offence of unlawful attack under Article 3 of the 1993 ICTY Statute] has been met, it should be borne in mind that commanders deciding on an attack have duties:

...

(b) to take all practicable precautions in the choice of methods and means of warfare with a view to avoiding or, in any event to minimizing, incidental civilian casualties or civilian property damage" (ICRC, n.d.b).

In its judgment in Ergi v. Turkey in 1998, the European Court of Human Rights held:

"The responsibility of the State is not confined to circumstances where there is significant evidence that misdirected fire from agents of the State has killed a civilian. It may also be engaged where they fail to take all feasible precautions in the choice of means and methods of a security operation mounted against an opposing group with a view to avoiding and, in any event, to minimizing, incidental loss of civilian life" (ICRC, n.d.b).

[20] This book's author agrees with the authors from a moral perspective as we hold a deontological view that some decisions should not be transferred to machines.

might be interpreted as a breach if it is proven that AWS are the more precise means/method of warfare.

However, we disagree with this perspective from an ethical standpoint and suggest in Part III, *de lege ferenda* considering the new ethical and legal challenges posed by autonomy, that States should not be held responsible for not using AWS.

Significantly enough, insightfully downplaying these pro-AWS consequentialist arguments, Amoroso highlights that "complete ouster of the human element from the targeting process would rule out a feasible precaution capable of preventing infringements of the principles of distinction and proportionality, as well as the right to life under IHRL" (Amoroso, 2020, p. 107). The decision-making of autonomous devices is qualitatively different from that of humans. Therefore, they make mistakes in different venues, and even the most technologically advanced AWS could generate causalities that would have been avoided had a human being been involved in the decision-making process (Amoroso, 2020, p. 108). Amoroso argues for an essential human fail-safe role in the process. "The obligation to do "everything feasible" to take lawful targeting decisions is best adhered to if States conceive of human-machine interactions in a way that ensure that the (distinct) weaknesses of humans and machines as reciprocally compensated" (Amoroso, 2020, p. 222). We agree with Amoroso that there shall be no absolute preference for machines or humans (Amoroso, 2020, p. 114). However, *de lege lata*, if it is proven in the concrete case that an AWS is the more precise means, in our view, States can be held responsible. It is necessary *de lege ferenda* to discuss the human-machine relation threshold, which was discussed in 7.1 and will be dealt with in Part III and also to bar a responsibility for not using AWS.

7.3 Weapons Review and State Responsibility

The obligation to review weapons is stated in article 36 of Additional Protocol I of 1977 and sets forth that States must "In the study, development, acquisition or adoption of a new weapon, means and methods of warfare"[21] assess if its use would "in some or all circumstances be prohibited by international law" (Additional Protocol I, 1977).

According to the ICRC's guide on weapons review, all States are obliged to review weapons, regardless of whether they are parties to the additional protocol or not, since States can only deploy weapons, means, and methods of warfare in a legal manner (ICRC, 2006).[22] Despite article 36's obligation not being widely

21 "The words "methods and means" include weapons in the widest sense, as well as the way in which they are used. The use that is made of a weapon can be unlawful in itself, or it can be unlawful only under certain conditions" (Pilloud and others, 1987, p. 398).

22 ICRC's guide to weapon review states that "The requirement that the legality of all new weapons, means and methods of warfare be systematically assessed is arguably one that applies to all States, regardless of whether or not they are party to Additional Protocol I. It flows logically from the truism that States are prohibited from using illegal weapons, means and methods of warfare or from using weapons, means and methods of warfare in an illegal manner. The faithful and responsible application of its international law obligations would require a State to ensure

(*Contd.*)

recognized as a rule of customary IHL,[23] in our view, at least its core aspect, which is a pre-deployment assessment of the legality of the weapon, is customary international law (Jevglevskaja, 2018).[24] In this sense are the Harvard Manual (HPCR, 2010, p. 9)[25] and Tallinn Manual 2.0 (Schmitt, 2017).[26]

A weakness of weapon review is the lack of an international normative framework (Amoroso, 2020, p. 253). To review weapons, States have to institute domestic procedures to determine their legality or not (Pilloud and others, 1987). Since States are free to determine those procedures (CCW GGE, 2019a),[27] they

---

that the new weapons, means and methods of warfare it develops or acquires will not violate these obligations. 6 Carrying out legal reviews of new weapons is of particular importance today in light of the rapid development of new weapons technologies" (ICRC, 2006).

23 The US for instance does not recognize it as customary international law and review weapons on a policy basis.

24 Jevglevskaja states about Article 36 of Additional Protocol I (AP I) "States, including military powers such as the United Kingdom, Germany, and France are known to have a weapons review procedure in place. Several militarily significant non-AP I States are not known to have instituted weapons review mechanisms. These States are India, Indonesia, Iran, Malaysia, Pakistan, Singapore, and Turkey. Only two States not formally bound by AP I—the United States and Israel—systematically carry out weapons reviews. Thus, even if the timeframe of nearly four decades since the adoption of AP I potentially satisfies the requirement that practice has to be pursued over a sufficient period of time, the insignificant number of States that are known to conduct weapons reviews render any argument in favor of "extensive and virtually uniform" State practice impossible" (Jevglevskaja, 2018, p. 209).
Nonetheless, she notes that the core obligation of a pre use assessment is required by customary international law: "Having found that Article 36 in its entirety has not crystalized into CIL, it remains to be examined whether the core aspect of the provision—the obligation to perform a pre-deployment analysis of weapons to determine their compliance with the LOAC—may have. Two expert publications, the Harvard Manual and Tallinn Manual 2.0, both suggest that this core aspect of Article 36 is CIL. A result of extensive, methodical and comprehensive reflection by many subject-matter experts on the existing rules of international law applicable to air and missile warfare, as well as cyber warfare, these manuals are based on the premise that failing to review new weapons prior to their deployment risks non-compliance with international law. Consequently, a pre-deployment legal analysis of a weapon is required as a matter of law" (Jevglevskaja, 2018, pp. 213–214).

25 The Manual on International Law Applicable to Air and Missile Warfare affirms "9. States are obligated to assess the legality of weapons before fielding them in order to determine whether their employment would, in some or all circumstances, be prohibited" (HPCR, 2010, p. 9).

26 According to the Tallinn Manual 2.0 on the International Law Applicable to Cyber Operations: "Rule 110 (a) All States are required to ensure that the cyber means of warfare that they acquire or use comply with the rules of the law of armed conflict that bind them. (b) States that are Parties to Additional Protocol I are required in the study, development, acquisition, or adoption of a new means or method of cyber warfare to determine whether its employment would, in some or all circumstances, be prohibited by that Protocol or by any other rule of international law applicable to them" (Schmitt, 2017).

27 The GGE CCW report states that "(i) Legal reviews, at the national level, in the study, development, acquisition or adoption of a new weapon, means or method of warfare are a useful tool to assess nationally whether potential weapons systems based on emerging technologies in the area of lethal autonomous weapons systems would be prohibited by any rule of international law applicable to that State in all or some circumstances. States are free to independently determine the means to conduct legal reviews although the voluntary exchange of best practices could be beneficial, bearing in mind national security considerations or commercial restrictions on proprietary information" (CCW GGE, 2019a).

might conclude differently if a weapon is legal or not, which encompasses a danger of fragmentation. The exchange of best practices is highly advisable but is not likely to resolve fragmentation (Amoroso, 2020, p. 253).

AWS review adds new challenges. Considering that environmental *stimuli* play a vital role in determining AWS actions, they will require testing in diverse scenarios and only be allowed to operate in the field under circumstances similar to testing (Crootof, 2018).[28] However, simulations do not accurately replicate circumstances on the ground, and it is impossible to anticipate all possible inputs and outputs to review legality (ICRC, 2019b, p. 3).[29] Moreover, there is a lack of previous cases to provide a framework for analyzing weapons capable of self-learning on the ground. This section's scope is to analyze whether the weapons review obligation is sufficient to deal with AWS.

7.3.1 Weapon Review: GGE's Guiding Principles and States' Perspectives

Three of the eleven Guiding Principles the GGE on AWS agreed on are weapon review related. Principle D emphasizes that "(...) in the study, development, acquisition, or adoption of a new weapon, means or method of warfare, determination must be made whether its employment would, in some or all circumstances, be prohibited by international law." Principle E emphasizes

28 Crootof highlights that "Given that its actions will be determined in part by stimuli from its environment, an autonomous weapon system will need to be tested in a variety of different virtual and real scenarios and only cleared for use in sufficiently similar real world circumstances. (...) Second, there is no precedent for evaluating weapon systems with the capacity for in-field learning. Traditionally, legal reviews presume that any significant change in software or hardware will be identified and evaluated before the altered weapon is fielded. But weapon systems with certain kinds of artificial intelligence might be able to evolve in response to environmental stimuli, such that a fielded autonomous weapon system will eventually be significantly different from the system originally approved. This problem might be addressed in various ways: by short-term deployments, by regularly scheduled evaluations and reboots, or by completely prohibiting in-field learning (at least with regard to target selection and engagement tasks). All of these solutions, however, would likely require new domestic guidelines, if not international standards(...) A silver lining is that the international conversation about how best to regulate autonomous weapon systems has the potential to highlight these issues and contribute to the development of widely-accepted standards for conducting legal reviews. Already, a few states have been more transparent about their internal review procedures in discussions of regulating autonomous weapon systems" (Crootof, 2018, pp. 64–67).

29 According to ICRC, "However, it is not possible to test all the potential inputs and outputs of the system for all circumstances, or even to know what percentage of the possible outputs one has tested. This means that it is difficult to formally verify and validate the predictability of the system and its reliability, or probability of failure. Considering weapon systems, it is therefore difficult to ensure that an autonomous weapon system is capable of being used in compliance with international humanitarian law, especially if the system incorporates AI – and particularly machine learning – control software. The more complex the environment, the more acute the problem of verification and validation. Given the limits of testing in the real world, computer simulations are used to increase the number of scenarios that can be tested. However, simulations bring their own difficulties, as building an accurate simulation is difficult and requires knowledge of all critical scenarios and how to re-create them faithfully. Simulations cannot generally replicate the real-world environment, even for simple tasks (ICRC, 2019b, p. 3).

that States that are acquiring or developing AWS must consider "(...) physical security, appropriate non-physical safeguards (including cyber security against hacking or data spoofing), the risk of acquisition by terrorist groups and the risk of proliferation (...)." Finally, Principle F states, "Risk assessments and mitigation measures should be part of the design, development, testing and deployment cycle of emerging technologies in any weapons systems" (GGE CCW, 2018). Considering that GGE's decisions are consensus-based, the three principles demonstrate that States consider weapons review of AWS extremely important.

To conduct weapons reviews, States must consider the specific features of emerging technologies in the context of AWS (Chair of the 2020 GGE, 2020).[30] In this sense, some delegations at CCW GGE stressed: "(...) the unique nature of LAWS, including potential for self-learning and associated unpredictability, that could pose novel challenges to weapons review processes" (CCW GGE, 2019a). Some States also highlighted the importance of "(...) greater transparency in the conduct of weapons reviews, as well as information-sharing on best practices" (CCW GGE, 2019a).

Focusing on sharing good practices, Australia, Canada, Japan, the Republic of Korea, the United Kingdom, and the US argue that a review of AWS shall be "conducted with an appropriate understanding of the weapons' capabilities and limitations, its planned uses, and its anticipated effects in those circumstances." One has to consider if the weapon is capable of being used in accordance with IHL, including "(...) whether the weapon is of a nature to cause superfluous injury or unnecessary suffering, or if it is inherently indiscriminate (...)", and also consider if the AWS use falls under specific rules, including CCW Protocols. The review process shall also advise possible measures that promote AWS compliance with IHL (Australia and others, 2022).

The UK highlighted that State responsibility arises from weapons review (Acheson, 2021, p. 17) and in the realm of sharing good practices, proposed a manual on AWS review and presented the following questions:

> "what information and level of understanding is necessary to inform an effective weapons review? Does the inclusion of AI functionality make a difference in relation to the scope of the review or the resulting authorisation? Does machine learning necessitate re-review and authorisation? If so how is this built into the review process and operationalised to ensure that the system does not exceed authorities? How should the approval parameters applied to the system as a result of a weapons review be best effected through the authorisations process and through into Rules of Engagement (RoE)?" (UK Delegation, 2022).

[30] The paper states that "In conducting legal weapons reviews, which are a legal obligation for parties to Additional Protocol I to the Geneva Conventions, governments must pay close attention to the particularities of emerging technologies in the area of lethal autonomous weapons systems" (Chair of the 2020 GGE, 2020).

The US, as already mentioned, also envisions that States should not aim at codification but to "exchange practice and implement holistic, proactive review processes that are guided by the fundamental principles of the law of war" (US Delegation, 2018a). The US also claimed the importance of the issue of human-machine interface in the process of weapons review (US Delegation, 2018a).[31]

In 2012, US DoD Directive 3000.09 assigned a responsibility that "(...) establishes requirements for verification and validation and test and evaluation" of autonomous and semi-autonomous weapons (US Delegation, 2018a). According to the US statement at GGE, that explains DOD directive 3009 requirements on AWS review:

> "Before fielding systems that would use autonomy in novel ways, such reviews must "assess system performance, capability, reliability, effectiveness, and suitability under realistic conditions, including possible adversary actions, consistent with the potential consequences of an unintended engagement or loss of control of the system." Such testing should include "analysis of unanticipated emergent behavior resulting from the effects of complex operational environments on autonomous or semi-autonomous systems." DoD Directive 3000.09 also requires that "safeties, anti-tamper mechanisms, and information assurance" have been implemented in autonomous and semi-autonomous weapon systems. These measures are intended to "minimize the probability or consequences of failures that could lead to unintended engagements or to loss of control of the system" by, for example, safeguarding against attempts by unauthorized individuals to fire the weapon" (US Delegation, 2018a, p. 2).

Despite not recognizing article 36 as a customary rule of IHL, the US reviews weapons on policy grounds (Dunlap, 2016, p. 65; US DoD, 2015).[32] In this

[31] The United States Delegation wrote that "32. We also recommend a proactive approach in addressing issues in human-machine interaction. States seeking to develop new uses for autonomy in their weapons should be affirmatively seeking to identify and address these issues in their respective processes for managing the life cycle of such weapons. For example, DoD Directive 3000.09 requires senior officials to review weapon systems that use autonomy in new ways. Such reviews, which are required before a system enters formal development and, again, before fielding, ensure that military, acquisition, legal, and policy expertise is brought to bear before new types of weapons systems are used. 33. This practice in conducting a special policy review is consistent with broader DoD practice in conducting legal reviews of the intended acquisition or procurement of any weapon by the Department of Defense, as reflected in US Department of Defense Directive 5000.01, The Defense Acquisition System. Such reviews, among other things, help ensure consistency with the law of war" (US Delegation, 2018a).

[32] The US Law of War Manual provides the following: "6.2.2 Questions Considered in the Legal Review of Weapons for Consistency With US Law of War Obligations. The review of the acquisition or procurement of a weapon for consistency with US law of war obligations should consider three questions to determine whether the weapon's acquisition or procurement is prohibited: • whether the weapon's intended use is calculated to cause superfluous injury; • whether the weapon is inherently indiscriminate; and • whether the weapon falls within a class of weapons that has been specifically prohibited. If the weapon is not prohibited, the review should also consider whether there are legal restrictions on the weapon's use that are specific to

(*Contd.*)

regard, the novel 2023 DoD Directive 3000.09, which reissues and cancels DoD Directive 3000.09 of 2012, has an entire section on guidelines for the review of certain AWS (US DoD, 2023).

Israel is also among the States claiming to share good practices. In Israel's view, those practices should be flexible and embrace diverse domestic models "so that Parties that have yet to establish their own national procedures can take inspiration from the various available models of legal review and adopt variations that best suit their own national systems" (Israel Permanent Mission to the UN Geneve, 2019a). Israel is not a party to the First Additional Protocol to the Geneva Conventions, but "(...) is of the view that applying legal reviews to new weapons is the best instrument for a State to ensure that it uses only lawful means of warfare during armed conflicts" (Israel Permanent Mission to the UN Geneve, 2016). Israel recognized the challenges in reviewing AWS and affirmed that States that have resources and aim to develop AWS also have to have adequate resources to review the weapons and ensure their predictability (Israel Permanent Mission to the UN Geneve, 2016).[33]

Unlike States such as the US, Russia considers the norm stated in article 36 as customary international law. According to GGE's on LAWS Principle E, the Russian Federation reviews weapons "within the framework of national procedures that help, based on the existing legal and regulatory framework, ensure proper control over compliance with the requirements of Article 36 of the AP I" (Russian Federation Delegation, 2020)[34]. They also emphasized that article 36 "(…) has no provisions on how exactly legal reviews should be conducted, and does not

---

that type of weapon. If any specific restrictions apply, then the intended concept of employment of the weapon should be reviewed for consistency with those restrictions" (US DoD, 2015).

33 The Israel Permanent Mission to the UN in Geneva stated that "While true that testing of sophisticated weapon systems, including systems with autonomous characteristics, could require more resources, the end goal is identical – the weapon should be reliable and its operation should be predictable. It is presumable that States who plan to develop LAWS would also have the resources to test them and ensure they are predictable."(Israel Permanent Mission to the UN Geneve, 2016)

34 The Russian Federation Delegation wrote that Article 7 of the Federal Law No. 275-FZ dated 29 December 2012 'On the State Defense Order' includes, among the main responsibilities of the state customer, the organization and conduct of tests of prototypes (complexes, systems) of weapons, military and special hardware, military property, preparation of documents for their adoption, as well as approval of technical documentation required for their development and mass production. Prototypes are assessed for such characteristics as distinction, 'no-excessive-damage,' etc., which should guarantee potential compliance of future weapons with IHL norms. In the context of the implementation of the guiding principles, of particular importance is the National Strategy for the Development of Artificial Intelligence (AI) for the period until 2030 adopted in 2019. As for principles g) and i), the Strategy sets out the need to identify and prevent any risks associated with the development of emerging technologies. In particular, it concerns the "inadmissibility of using AI for the purpose of deliberate infliction of harm to individuals and legal entities, as well as prevention and minimization of risks of negative consequences of using AI technologies" (paragraph 19B). It emphasizes that "the creation of universal AI, similar to a human being, can lead to negative consequences due to social and technological changes that accompany the development of AI technologies (paragraph 9)" (Russian Federation Delegation, 2020).

impose an obligation on States to make their results public, nor to provide anyone with information on the subject" (Russian Federation Delegation, 2020).

France reinforced the need for a national procedure. To operationalize Guiding Principle E and according to international obligations, especially article 36 of the Additional Protocol I to the Geneva Conventions, States must enact "(…) a national procedure to review the legality of weapons systems that they develop or acquire, including those based on emerging technologies in the area of LAWS (…)" (French Delegation, 2020).

China affirmed that State national weapons reviews positively impact the prevention of "(…) the misuse of relevant technologies and on reducing harm to civilians" (People's Republic of China Delegation, 2018). Nonetheless, exposing a problem with France's proposal, China highlighted that national regulations and practices vary significantly: "There is much uncertainty concerning the results of these reviews and how the results are treated. Therefore, any initiative or proposal based on such reviews can hardly solve, in a fundamental way, the concerns caused by LAWS" (People's Republic of China Delegation, 2018).

Switzerland acknowledged the specific challenge AWS pose to weapons reviews: "Specifically, the question is how such systems and their specific characteristics can be meaningfully tested." The Swiss delegation highlighted that there are not only technical challenges for verifying AWS compliance with IHL, but "(…)there is also a conceptual challenge related to the fact that an autonomous system will assume an increasing number of determinations in the targeting cycle which traditionally are being taken care of by a human operator" (Switzerland Delegation, 2017). They exemplify that if assessing proportionality is transferred from the human operator (in traditional weapons) to AWS, the review must also verify the weapon's proportionality assessment. Switzerland claimed that those conceptual challenges might require novel procedures to review AWS. Examples of measures that the national review process should take are mechanisms of human override of the weapon in case of malfunction and in-depth knowledge about the system's predictability while interacting with other autonomous devices (Switzerland Delegation, 2017).

Argentina, Austria, Belgium, Chile, Costa Rica, Ecuador, Guatemala, Ireland, Kazakhstan, Liechtenstein, Luxembourg, Malta, Mexico, New Zealand, Nigeria, Panama, Peru, the Philippines, Sierra Leone, Sri Lanka, the State of Palestine, Switzerland, and Uruguay acknowledged the challenges and value of weapons reviews. It is insufficient "to deal with all issues autonomous weapons systems raise" (Argentina and others, 2022d). At the same time, "weapon reviews play an important complementary role and there is value in strengthening such reviews" (Argentina and others, 2022d).

Calling for a legally binding instrument, Draft Protocol VI submitted by Argentina, Ecuador, Costa Rica, Nigeria, Panama, the Philippines, Sierra Leone, and Uruguay foresees weapons reviews in its article 4.

Sec 1: Each High Contracting Party shall ensure that weapon systems under development or modification which changes the effects or use of existing weapon systems, including as a result of self-learning process, must be reviewed to ensure compliance with international law.

Sec 2: Each High Contracting Party shall be transparent regarding all aspects of the development of autonomous weapon systems across their entire life cycle, including national processes for reviewing them, taking into account the system's self-learning capabilities (...)" (Argentina and others, 2022b).

Chile and Mexico argued that AWS legal review:

"must include an assessment that allows for the understanding of the attributes and effects in weapons with autonomous capabilities, as well as its conformity with international humanitarian law and international law, in particular:

1. Evaluate its technical performance, including in terms of reliability and predictability and whether its foreseeable effects are capable of being limited to military objectives and controlled in time and space;
2. Confirm its intended or expected use; and
3. Confirm the placement of adequate limits on tasks and types of targets.

Legal reviews of weapons autonomous functionalities should adopt a precautionary approach and deny authorization when there might be less than full certainty of all the characteristics listed in the paragraph above" (Chile and Mexico Delegations, 2022).

In short, while some States claim to share good practices, others claim for legally binding instruments. While some stay at a broader legality assessment, others propose concrete measures, but all of them seem to recognize the challenges and relevance of AWS review.

7.3.2 Back to the Question: Weapon Review and AWS

The focus of the present book is State responsibility. We argue that *de lege lata* the existing obligation to review weapons applies to AWS as it does to any other weapons (under article 36 of 1977 Additional Protocol I to the Geneva Conventions of 1949 and its core aspects under customary international law), and also based on the GGE Principles D, E and F which were unanimously agreed upon being, thus, soft law. Therefore, weapons review obligation is a relevant source of State responsibility. In fact, "(...) adequate testing and reviews may also have implications on the level of State responsibility, including for malfunction of approved autonomous weapon systems" (Switzerland Delegation, 2017). If the obligation to review weapons, which is a State duty, is ascertained and violated, they can be held responsible. In this case, it is clear-cut that the State is at fault. Thus, there are no attribution challenges in the realm of weapons review, as there are roadblocks regarding acts committed by AWS. Furthermore, most of AWS responsibility challenges presented in the former chapters are not an issue in the AWS review.

> "What kind of legal review is required is a question States must address, regardless of what type of weapon you are talking about. AWS make the old ambiguities about what is required in a legal review more salient, in part because they raise the possibility of in-field learning. (…) They raise all kinds of new problems too, in terms of what should be included in a legal review (…) Consider the data question. What are a State's obligations, when training AWS on different data sets? That is getting into the weeds, but that's very relevant to how they perform in the field. And that question just hasn't been addressed. And is that limited to AWS? No, it's relevant to any AI system, particularly any AI predictive system. (…) Also, how do you test a system that is capable of learning from the test itself? (…)If States aren't willing to ban in-field learning, then you've also got to talk about weapons review for weapon systems with in-field learning capabilities. I do not think what we have right now is adequate to address that. There will need to be clarity about what constitutes a sufficient review of the training data, review of the testing practices, and review of the fielding practices. The law is already ambiguous, and autonomous weapons systems make it worse" (Crootof, personal communication, May 26, 2022).

In summary, despite being subject to existing IHL obligations, our perspective is that existing international obligations of weapons review during the "study, development, acquisition or adoption" of a "new weapon, means or method of warfare" does not suffice for AWS, and a tailored weapon review process for AWS would certainly enhance responsibility. We recall that the lack of an international framework on how to review weapons is a challenge for weapons review in general. The drawbacks of a lack of an international framework are more significant with AWS, since they encompass AI, might be capable of self-learning, and reviewing is even more challenging. GGE Guiding Principles D, E, and F point towards the need for a tailored review process for AWS. Furthermore, the majority of State statements above also evidence the need for a specific weapons review for AWS. Significantly enough, at CCW, States have supported "the need of weapon review if AWS alter significantly as regards its operational functioning" (Moyes, 2019, p. 7–8). This metamorphosis is a specificity of AWS, considering that the target profiles might change through machine learning. If such innovation exists, and the emerging target profiles were not previously approved, a new review or a constant review to monitor it is required. Otherwise, the purpose of weapons review would vanish.

> I do think that appropriate assurance for any kind of AI system, doing test and evaluation to make sure that it is reliable and robust in real-world operating conditions, is very important. There is not currently an international requirement to do that. Presumably, States don't want to see

> their weapons systems fail, but there are no international standards or best practices or norms and expectations for which States ought to be doing. So I think both would be valuable. I think that a treaty would probably be too onerous, but some degree of best practices or expectations among States about what should legal weapons reviews surrounding AI and autonomy in weapons include, and the expectation of testing and evaluating systems to ensure that they are reliable, would also be important. What should that include, what should be done if they malfunction and strike the wrong targets? Both of those are important and I think there's value in the international community trying to better understand those and potentially reach an agreement on some set of standards or guidelines or best practices (Scharre, personal communication, May 27, 2022).

Thus, an adapted approach to the weapons review process is necessary to assess whether an AWS complies with IHL and IHRL (Sanders and Copeland, 2020). Prospectively, we argue for more than good practices, since we believe that the international community would benefit from international binding rules on reviewing AWS.

7.4 State Responsibility Due to Breach of the Due Diligence Obligation

In this section, the question is if, already *de lege lata*, a State can be held responsible for AWS misdoings due to the breach of due diligence. It also questions whether further rules specifying State due diligence obligations regarding AWS are necessary.

The ARSIWA does not contain the expression 'due diligence,' as they do not consider the issue of fault. Therefore, any due diligence concept is usually taken from other contexts (French and Stephens, 2014, p. 4–5).[35] However, the ILC commentary on the ARSIWA touches upon due diligence as regards activity by private actors by stressing that a "State may be responsible for the effects of the conduct of private parties if it failed to take necessary measures to prevent those effects" (ILC, 2002b, p. 39). This section will start with a general overview of due diligence and, next, analyze it through two perspectives: State breach of due diligence regarding a State's own acts and State breach of due diligence regarding activities by private actors.

Every State is obliged to forbid that its territory is used to knowingly violate the rights of other States (ICJ, 1949, p. 22). The "principle of prevention, as a customary rule, originates in the due diligence that is required of a State in its territory" (ICJ, 2010, par. 101). States bear the obligation to conduct themselves as a responsible subject, which means all efforts must be taken to prevent a breach of an international obligation, including adopting rules and measures, and the

[35] An example of those other contexts is the Draft Articles on the Prevention of Trans boundary Harm from Hazardous Activities.

obligation of vigilance, applicable to public and private actors (ICJ, 2010, p. 79). A State can be held responsible if it fails to accomplish this duty (Koivurova, 2010; French and Stephens, 2014, p. 2).[36]

The State responsibility due to a breach of due diligence varies according to the circumstances, the primary obligation's content, and the laws' object and purpose (ILC 2001b, p. 34).

> Among the factors that make such a description difficult is the fact that "due diligence" is a variable concept. It may change over time as measures considered sufficiently diligent at a certain moment may become not diligent enough in light, for instance, of new scientific or technological knowledge. It may also change in relation to the risks involved in the activity.... The standard of due diligence has to be more severe for the riskier activities (ITLOS, 2011, p. 43).

Note that it is an obligation of procedure and not of result. Therefore, if damage occurs, it does not necessarily mean a due diligence violation has occurred. It embraces both the State's obligations as well as the obligation of a State as regards actors that are not under its direct command or authority.

The two conditions for responsibility to ensure are: "(i) it had the means to prevent or to repress the unlawful act and (ii) it knew or should have known about the risk of the violation" (Berkes, 2018, p. 445). The means to prevent and repress may vary according to the circumstances, and its assessment includes the economic level of the State (ILC, 2001a, p. 155),[37] the "level of scientific knowledge and technical capability available to a given State in the relevant scientific and technical fields," (ITLOS, 2011, par 161–162)[38] its capacity to impact on

---

36 According to the Max Planck Encyclopedia of Public International Law: "Due diligence is an obligation of conduct on the part of a subject of law (Subjects of International Law). Normally, the criterion applied in assessing whether a subject has met that obligation is that of the responsible citizen or responsible government (Governments). Failure on a subject's part to comply with the standard—often termed negligence—describes the blameworthiness of the subject as one element of ascribing legal responsibility to it" (Koivurova, 2010).
See also the first Report Study Group on Due Diligence in International Law that states, "due diligence is concerned with supplying a standard of care against which fault can be assessed." "It is a standard of reasonableness, of reasonable care, that seeks to take account of the consequences of wrongful conduct and the extent to which such consequences could feasibly have been avoided by the State or international organization that either commissioned the relevant act or which omitted to prevent its occurrence" (French and Stephens, 2014, p. 2).

37 According to ILC's Draft Articles on Prevention of Trans boundary Harm from Hazardous Activities, "The economic level of States is one of the factors to be taken into account in determining whether a State has complied with its obligation of due diligence. But a State's economic level cannot be used to dispense the State from its obligation under the present articles" (ILC, 2001a, p. 155).

38 According to ITLOS, "161. As pointed out in paragraph 125, the provisions of the Nodules Regulations and the Sulphides Regulations that set out the obligation for the sponsoring State to apply a precautionary approach in ensuring effective protection of the marine environment refer to Principle 15 of the Rio Declaration. As mentioned earlier, Principle 15 provides that the precautionary approach shall be applied by States "according to their capabilities". It follows that the requirements for complying with the obligation to apply the precautionary approach

(*Contd.*)

the conducts on non-governmental agents (ICJ, 2007, p. 221; Berkes, 2018, p. 445),[39]and the risk of the activity (French and Stephens, 2016 p. 2; ITLOS, 2011, par. 117; ILC 2001a, par. 11; UNGA, 2011, prin. 17). The knowledge condition limits due diligence responsibility to damages that can be foreseen or anticipated (ILC, 2001a, p. 155),[40] which might be challenging in the context of LAWS, since it has a degree of unpredictability. Nonetheless, unpredictability, despite creating barriers, does not bar due diligence. These conditions will be further specified in the subsequent subheading that deals with the due diligence obligation of a State regarding its activities and private actors' activities.

The "Basic Principles and Guidelines on the Right to a Remedy and Reparation for Victims of Gross Violations of International Human Rights Law and Serious Violations of International Humanitarian Law" reinforces that States are obliged to "respect, ensure respect for and implement international human rights law and international humanitarian law as provided for under the respective bodies of law (...)" (UNGA, 2005). The document further specifies concrete obligations to ensure respect, such as enacting legislative and administrative measures to prevent breaches (UNGA, 2005). On those grounds, we claim that State failure to regulate AWS through legislative and administrative measures or monitor the activities of private agents might be a breach of due diligence and entail responsibility.[41]

---

may be stricter for the developed than for the developing sponsoring States. The reference to different capabilities in the Rio Declaration does not, however, apply to the obligation to follow "best environmental practices" set out, as mentioned above, in regulation 33, paragraph 2, of the Sulphides Regulations.

162. Furthermore, the reference to "capabilities" is only a broad and imprecise reference to the differences in developed and developing States. What counts in a specific situation is the level of scientific knowledge and technical capability available to a given State in the relevant scientific and technical fields" (ITLOS, 2011, par 161–162).

The opinion was issued in the field of environmental law, but the reasoning can be applied mutatis mutanda also to LAWS.

39 According to the ICJ: "In this area the notion of 'due diligence,' which calls for an assessment in concrete, is of critical importance. Various parameters operate when assessing whether a State has duly discharged the obligation concerned. The first, which varies greatly from one State to another, is clearly the capacity to influence effectively the action of persons likely to commit, or already committing, genocide. This capacity itself depends, among other things, on the geographical distance of the State concerned from the scene of the events, and on the strength of the political links, as well as links of all other kinds, between the authorities of that State and the main actors in the events. The State's capacity to influence must also be assessed by legal criteria, since it is clear that every State may only act within the limits permitted by international law; seen thus, a State's capacity to influence may vary depending on its particular legal position vis-à-vis the situations and persons facing the danger, or the reality, of genocide. On the other hand, it is irrelevant whether the State whose responsibility is in issue claims, or even proves, that even if it had employed all means reasonably at its disposal, they would not have sufficed to prevent the commission of genocide" (ICJ, 2007, p. 221, par 430).

40 According to ILC "The required degree of care is proportional to the degree of hazard involved. The degree of harm itself should be foreseeable and the State must know or should have known that the given activity has the risk of significant harm. The higher the degree of inadmissible harm, the greater would be the duty of care required to prevent it" (ILC, 2001a, p. 155 par. 18).

41 Barnidge states that "Like the due diligence principle generally under international law, whether

(*Contd.*)

In the field of IHL, Common Article 1 of the Geneva Conventions, considered customary international law, as stated in ICRC rule 144 (ICRC n.d.d),[42] foresees the obligation to respect and ensure respect for the Conventions, which is interpreted as a due diligence obligation. This obligation "is not limited to behavior by parties to a conflict, but includes the requirement that States do all in their power to ensure that international humanitarian law is respected universally" (ICRC n.d.d). However, it remains divisive regarding as to whose acts States bear this obligation (Berkes, 2018, p. 441) and what ensuring respect entails. In a narrow interpretation, it means the State's agents, territory, and jurisdiction. In a broader interpretation, to ensure respect has an external dimension and imposes a negative obligation: not to encourage, aid, or assist. In short, the boundaries of what conduct could trigger State responsibility are not yet defined.

This section aims to analyze to what extent States have a due diligence obligation to prevent misdoings in the context of AWS, meaning if States fail to do so, they can be held responsible for a breach of due diligence.

7.4.1 State Breach of Due Diligence as Regards its Own Activities, State-Sponsored Activities, and Private Actors' Activities

Aspects of the State's due diligence obligation regarding AWS, such as weapons review and precautionary attack measures, were considered in the previous chapters. In this chapter, we will zoom in on the general aspects of due diligence as regards AWS.

A concrete example of State due diligence obligations in the context of AWS is to take constant care to refrain from targeting the civilian population and objects. "All feasible precautions must be taken to avoid, and in any event to minimize, incidental loss of civilian life, injury to civilians and damage to civilian objects" (UK Delegation, 2022). In the same sense, States must take precautions in attack and must "cancel or suspend an attack if it becomes apparent that the target is not a military objective or is subject to special protection, or that the attack may be expected to violate the rule of proportionality, as required by the rules on precautions in attack" (Davison, 2017).

Regarding private actors' activities, AWS require a high technology level, a crucial feature in this new warfare paradigm (Slijper, Beck, and DaanKayser, 2019). Thus, States need private-sector AI companies to be able to maintain leadership in the military field (Del Monte, 2018, p. 63).[43] Considering "(…) the

---

a state has complied with the obligations spelled out in General Assembly Resolution 60/147 will trigger an assessment on the basis of a reasonableness standard" (Barnidge, 2006).

42 ICRC customary IHL database in rule 144 foresees, "States may not encourage violations of international humanitarian law by parties to an armed conflict. They must exert their influence, to the degree possible, to stop violations of international humanitarian law" (ICRC n.d.d).

43 Louis del Monte affirms that "This means that if the US military intends to continue superiority in autonomous weapon technologies that require artificial intelligence, the United States will need to restructure its autonomous weapons procurements to include procurements from commercial corporations with AI expertise" (Del Monte, 2018, p. 63).

extraordinarily demanding tasks they will be required to perform, one can easily foresee that AWS will develop as a complex "system of systems," programmed and manufactured by joint ventures of private and public companies" (Amoroso, 2020, p. 129). DoDs, to keep up with innovative technology, are developing partnerships with the private sector or contracting private agents (Finland, Estonia, France, Germany, and the Netherlands delegations, 2019, p. 1).[44] China is using private-sector technology to develop its next-generation warfare (Allen, 2019). The US also highlighted its strategic defense documents and the necessity to collaborate with the private sector (Slijper, Beck and Kayser, 2019).[45]

When the breach cannot be directly attributed to the State, as in the case of public-private partnerships or activity by private actors, based on the principle of due diligence, States might be held responsible for their actions or omissions in the context of AWS breaches by private actors.[46]

We recall that the ARSIWA avoids a broad attribution approach grounded solely on a link to the State, such as nationality, habitual residence, or incorporation. The aim is both: to limit State responsibility and to ensure the autonomy of individual actors. The ground rule is that the international responsibility of the State must involve wrongful acts tied to the State as an organization and that only if the conduct can be attributed to the State, according to international law, can it be deemed the conduct of the State. Notably, the root of attribution is international law and not merely on a bond of factual causality. Attributing an act to a State is a distinct operation from characterizing the conduct as wrongful under international law (ILC, 2001b, pp. 38–39).[47]

---

44 (Finland, Estonia, France, Germany, and the Netherlands delegations, 2019, p. 1) The food for thought paper affirms that "Traditional military stakeholders and prime contractors are challenged by cutting-edge research and development in the private sector, including small and medium-sized enterprises and non-traditional contractors. Given that, these technologies are mainly produced in the commercial civil sector, and noting the remarkable investments of countries like US and China, Europe must keep up with the competition" (Finland, Estonia, France, Germany, and the Netherlands delegations, 2019, p. 1).

45 PAX for Peace Institute report highlights that "Acknowledging the innovative power of the private sector, the DoD is keen to have better connections with the engineers in Silicon Valley. Indeed, recent initiatives demonstrate that public-private partnership is a US military AI priority. One such initiative is the Defense Innovation Unit Experimental (DIUx), set up in 2015 and "meant to serve as a liaison between the Defence Department and the tech world". The DIUx contracts companies "offering solutions in a variety of areas—from autonomy and AI to human systems, IT, and space—to solve a host of defence problems". The DIUx was set up initially as an experiment, but in August 2018 the DoD announced that it would be renamed the Defense Innovation Unit (DIU) "to convey a sense of permanence to the agency" (Slijper, Beck and Kayser, 2019).

46 In a similar vein, Barnidge, 2006 states, in the context of terrorism, "In an age of modern terrorism, for example, the due diligence principle can be used to hold states responsible for their actions or omissions when dealing with international terrorism committed by non-state actors. How, whether, and with what breadth the due diligence principle is applied, however, remains one of political will" (Barnidge, 2006).

47 ILC commentary is that "(2) In theory, the conduct of all human beings, corporations or collectivities linked to the State by nationality, habitual residence or incorporation might be attributed to the State, whether or not they have any connection to the Government. In

(*Contd.*)

Even when the breaches themselves are not attributed to States, because they were private actors' activities, States must act with due diligence as regards those private actors' activities. If private actors commit wrongful acts, States might be held responsible if they fail to prevent and repress those breaches. This duty is grounded on the customary principle of prevention (ICJ, 2010, par. 101).[48] Thus, if the conduct is of a private actor, or even if the State-private actor's partnership does not imply the exercise of State power or the State being in command or authority, responsibility can still be attributed to States due to the breach of due diligence. Private actors also bear a due diligence obligation (UNGA Human Rights Council, 2011),[49] but it will not be discussed here since our focus is on State responsibility.

As mentioned in the previous section, the two conditions for due diligence responsibility to ensure are: "(i) it had the means to prevent or to repress the

---

international law, such an approach is avoided, both with a view to limiting responsibility to conduct which engages the State as an organization, and also so as to recognize the autonomy of persons acting on their own account and not at the instigation of a public authority. Thus, the general rule is that the only conduct attributed to the State at the international level is that of its organs of government, or of others who have acted under the direction, instigation or control of those organs, i.e. as agents of the State (…) (4) The attribution of conduct to the State as a subject of international law is based on criteria determined by international law and not on the mere recognition of a link of factual causality. As a normative operation, attribution must be clearly distinguished from the characterization of conduct as internationally wrongful. Its concern is to establish that there is an act of the State for the purposes of responsibility. To show that conduct is attributable to the State says nothing, as such, about the legality or otherwise of that conduct, and rules of attribution should not be formulated in terms which imply otherwise. But the different rules of attribution stated in chapter II have a cumulative effect, such that a State may be responsible for the effects of the conduct of private parties, if it failed to take necessary measures to prevent those effects" (ILC, 2001b, pp. 38–39).

48 According to the ICJ: "101. The Court points out that the principle of prevention, as a customary rule, has its origins in the due diligence that is required of a State in its territory. It is "every State's obligation not to allow knowingly its territory to be used for acts contrary to the rights of other States" (Corfu Channel (United Kingdom v. Albania), Merits, Judgment, I.C.J. Reports 1949, p. 22). A State is thus obliged to use all the means at its disposal in order to avoid activities which take place in its territory, or in any area under its jurisdiction, causing significant damage to the environment of another State. This Court has established that this obligation "is now part of the corpus of international law relating to the environment" (Legality of the Threat or Use of Nuclear Weapons, Advisory Opinion, I.C.J. Reports 1996 (I), p. 242, para. 29)" (ICJ, 2010, par. 101).

49 According to the Guiding Principles on Business and Human Rights:
"17. In order to identify, prevent, mitigate and account for how they address their adverse human rights impacts, business enterprises should carry out human rights due diligence. The process should include assessing actual and potential human rights impacts, integrating and acting upon the findings, tracking responses, and communicating how impacts are addressed. Human rights due diligence:
(a) Should cover adverse human rights impacts that the business enterprise may cause or contribute to through its own activities, or which may be directly linked to its operations, products or services by its business relationships;
(b) Will vary in complexity with the size of the business enterprise, the risk of severe human rights impacts, and the nature and context of its operations;
(c) Should be ongoing, recognizing that the human rights risks may change over time as the business enterprise's operations and operating context evolve" (UNGA Human Rights Council, 2011, p. 16, principle 17).

unlawful act and (ii) it knew or should have known about the risk of the violation" (Berkes, 2018, p. 445).

According to the ILA study group report, "There are some obligations of due diligence for which the same standard is required of all States and no account is taken of the economic resources of a State, its degree of control over its territory or other special characteristics." Among those, it cites the prevention of terrorism and IHL (French and Stephens, 2016, p. 18). Also, according to the ILA study group report, IHL due diligence standards vary depending on "the extent of control that parties have over territory, whether effective or overall control, as well as the role of non state parties, including private military contractors and international organizations" (French and Stephens, 2016, p. 15).

We claim that at the helm of LAWS, for the State's own acts—the State deploying AWS—or States sponsoring AWS activities or private acts, it seems reasonable to require, as a general rule, the same level of due diligence from States considering the high risks AWS pose, the military advantages gained by States, and that they aimed at targeting using an inherently risky activity, and also that most of the breaches occur in the framework of IHL. The exception is that due diligence shall vary depending on the parties' control over the territory or the role of non-State actors in the process. However, this could be interpreted differently; therefore, as will be explained in Part III, a specific statement *de lege ferenda* is advisable.

As regards the second requisite, the State's knowledge, we acknowledge the challenges since AWS are inherently unpredictable. However, States that deploy AWS are required throughout the entire cycle of the AWS to test, evaluate, monitor, and anticipate possible damages. It is within this knowledge framework that responsibility for due diligence ensues. If this is not the case, there is no breach of the due diligence obligation. However, this could also be interpreted differently; therefore, as will be shown in Part III, a specific statement *de lege ferenda* is advisable.

Regarding State knowledge of private actors' activities, "in some circumstances a State may be under a specific obligation to use best efforts to gain knowledge of activity within its territory or jurisdiction" (French and Stephens, 2016, p. 11). Furthermore, States must monitor some activities taking place in their territory (ICJ, 1949, p. 18), and the development and use of LAWS must undoubtedly be one of them, as States are obliged, for instance, to review weapons (the specific issue of weapons review was dealt with in subheading 7.3). Furthermore, for State-sponsored activities, States are expected to act with due diligence even if there is insufficient scientific evidence of the risks, but there are plausible indicators thereof (ITLOS, 2011, par. 131).[50] However, this interpretation is subject to diverse views and requires clarification *de lege ferenda*.

[50] According to ITLOS: "131. Having established that under the Nodules Regulations and the Sulphides Regulations, both sponsoring States and the Authority are under an obligation to apply the precautionary approach in respect of activities in the Area, it is appropriate to point out that the precautionary approach is also an integral part of the general obligation of due diligence of sponsoring States,

(*Contd.*)

7.4.2 Back to the Question: States due Diligence Obligation and AWS

Back to the question we pose in this section, the conclusion is that already *de lege lata*, a State can be held responsible for AWS misdoings due to the breach of due diligence. For example, State failure to regulate AWS through legislative and administrative measures or monitor the activities of private agents might be a breach of due diligence and entail responsibility (Barnidge, 2006).[51]

Also corroborating the State due diligence responsibility, are the Principles and Good Practices on Emerging Technologies in the Area of Lethal Autonomous Weapons Systems - Proposed by Australia, Canada, Japan, the Republic of Korea, the United Kingdom, and the United States:

> During the design, development, testing, and deployment of weapons systems based on emerging technologies in the area of LAWS, the risks, *interalia,* of civilian casualties, as well as precautions to help minimize the risk of incidental loss of life, injuries to civilians, and damage to civilian objects must be considered. Other types of risks should be considered, as appropriate, including but not limited to the risk of unintended engagements, risk of loss of control of the system, risk of proliferation, and risk of acquisition by terrorist groups. (Australia and others, 2022).

However, due diligence is not specific in terms of what exactly is required (Bruun, personal communication, June 21, 2022),[52] and its contours in the context of AWS are yet to be defined by States, scholars, and courts. We offered an interpretation as regards the requisite of "knew or should have known about the risk of the violation" (Berkes, 2018, p. 445), but they could be interpreted differently. Further rules specifying State due obligations diligence as regards AWS are necessary to elucidate issues such as the standard required from States

---

which is applicable even outside the scope of the Regulations. The due diligence obligation of the sponsoring States requires them to take all appropriate measures to prevent damage that might result from the activities of contractors that they sponsor. This obligation applies in situations where scientific evidence concerning the scope and potential negative impact of the activity in question is insufficient but where there are plausible indications of potential risks. A sponsoring State would not meet its obligation of due diligence if it disregarded those risks. Such disregard would amount to a failure to comply with the precautionary approach" (ITLOS, 2011, par. 131).

51 According to Barnidge, "Like the due diligence principle generally under international law, whether a state has complied with the obligations spelled out in General Assembly Resolution 60/147 will trigger an assessment on the basis of a reasonableness standard" (Barnidge, 2006).

52 This is a citation from the author's Ph.D. dissertation. During an interview for the author's Ph.D., Laura Bruun, in her personal capacity, stated: "State's responsibility applies not only for acts, but for omissions, and I think that it is all the things States are not going to do that will harm civilians in relation to AWS. All the due diligence obligations and precautionary measures that States have to take (In our work we talk about, just to give some examples, legal review, legal advice, training obligations, and instructions). All these obligations that States have to implement to ensure that in the end no one is harmed. All these obligations are not really specific in terms of what they require or when a legal review is sufficient" (Bruun, personal communication, June 21, 2022).

as regards the means to prevent the unlawful act. Considering AWS' inherent unpredictability, we argue that a lower threshold of "knew or should have known about the risk of the violation" shall apply. Therefore, we claim the necessity of establishing a due diligence framework regarding AWS. In Part III, which tackles possibilities *de lege ferenda*, we will discuss some possibilities, such as impact assessments and permit procedures for all AWS activities, and establish a requirement of monitoring AWS activities, inspired by the ILA report on the environmental law framework of due diligence and ICJ case-law (French and Stephens, 2014, p. 28; ICJ, 2015, p. 665, par. 153).

Part III

Mind the Gaps: Venues to Enhance State Responsibility and Accountability

The fundamental question of this book is whether the current regime on international responsibility of States based on the ARSIWA, which was drafted without considering AI, suffices to ensure State responsibility in cases of violations generated in the context of AWS. If not, what are the main gaps, and what are some possible approaches to address those gaps?

Throughout the chapters of Part II, we analyzed the ARSIWA as regards AWS misdoings and State responsibility and demonstrated that the current regime of international responsibility of States is insufficient to address AWS challenges. We found, in short, the following thirteen issues of concern, which, focusing on possible solutions *de lege ferenda* we grouped into three clusters: (1) Issues that can be addressed through tailored attribution rules; (2) Issues that can be addressed through a strict liability regime, and (3) Issues that can be addressed through other AWS specific rules. Some of the issues might fit into more than one cluster.

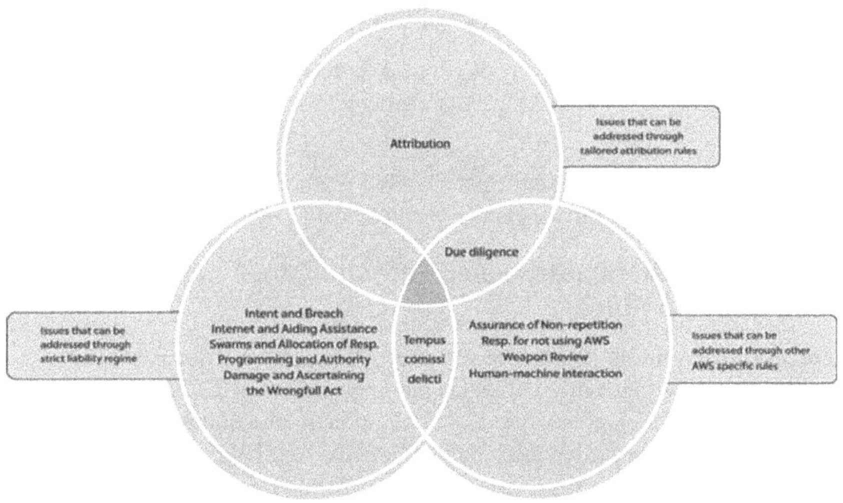

Issues that can be Addressed Through Tailored Attribution Rules

Attribution Issue: Considering that the ARSIWA is grounded on a human action paradigm, it is challenging to attribute responsibility to the State when the action or omission occurred in the context of AWS.

Due Diligence Issue: The breach of the duty of due diligence might be a ground to hold States responsible for their own acts and private acts, but the standard of due diligence as regards AWS needs to be established to enhance responsibility.

Issues that can be Addressed Through a Strict Liability Regime

Intent and Breach Issue: Rules of IHL and IHRL often require intent, which might hinder the very breach of an international obligation when AWS perform the conduct.

Intent and Aiding Assistance Issue: Considering that article 16 of the ARSIWA requires intentionality from the State's agents, responsibility for aiding assistance in the context of AWS might be challenging outside the framework of IHL.

Swarms and Allocation of Responsibility Issue: Allocation of responsibility in the case of swarms composed by AWS from different States is uncertain under the ARSIWA, especially if swarms decide on a consensus basis or through emergent coordination.

Programming and Authority Issue: Even if international law evolves to consider AWS and State organs, it is unclear if AWS programmed by a sending State can be deemed to be under the authority of a receiving State.

Damage and Ascertaining the Wrongful Act Issue: In the context of AWS and algorithmic decision-making, it is challenging to ascertain the wrongful act if there is no damage.

***Tempus Comissi Delicti* Issue:** Specific AWS features, such as opaqueness, add challenging roadblocks to assess the *tempus comissi delicti*.

Issues that can be Addressed Through Other AWS-specific Rules

***Force Majeure* Issue:** The mere fact that AWS may violate international law and that they are inherently unpredictable in general does not allow the defense argument of *force majeure* due to contribution to the situation. Nonetheless, States might present a good faith defense, and the ARSIWA might be interpreted differently.

Assurance of Non-repetition Issue: AI and AWS add further challenges to the assurance of non-repetition due to their inherent unpredictability.

Responsibility for not Using AWS Issue: *De lege lata*, not using AWS can be regarded as a violation if a State had the option of using AWS, and it was proven that such use would prevent a breach. However, *de lege ferenda*, considering the new ethical and legal challenges posed by autonomy, we argue against holding States responsible for not using AWS.

Weapons Review Issue: Despite being a relevant source of State responsibility, the existing obligation to review weapons needs to be adapted to address AWS better.

Human-machine Interaction Issue: International law requires human-machine interaction. Therefore, if a State fails to do so, it might be held responsible. However, the threshold of this interaction and the kind of responsibility it gives rise to must be established by the international community.

***Tempus Comissi Delicti* Issue** (which also fits in the second cluster)

Due Diligence Issue (which also fits in the first cluster)

Part III aims to think through the last part of the book's question, which concerns venues to enhance State responsibility and accountability regarding AWS misdoings, *de lege ferenda*

Which possible approaches might bridge those gaps?

Two important remarks. First, as mentioned in the methodological justification, the book focused *de lege lata* on the analyses of the international responsibility of States, as stated by the ARSIWA, since such a regime is consolidated in customary law and there is no general international liability regime in international law, nor a specific liability regime dealing with AWS. *De lege ferenda*, many of the proposed measures embrace responsibility and liability, which means a duty to compensate for damage even if there is no breach of an international obligation since AWS activities are inherently risky and afford a significant military advantage to the States that deploy them. Second, this book focuses on the substance, meaning the possible content of norms to address the challenges presented, rather than the form, which will be briefly dealt with in Chapter 9 for the sake of clarity.

The following chapters will first provide a short overview of the regulative options and then discuss scholarship on substantive possibilities *de lege ferenda* (Chapter 8), then will deal with each of the 3 clusters of issues raised. Chapter 9 covers issues that can be addressed through tailored attribution rules; Chapter 10 deals with issues that can be addressed through a strict liability regime; and Chapter 11 looks at issues that can be addressed through other AWS-specific rules.

CHAPTER

8

Autonomous Weapons Systems: Regulative Options and Substantive Possibilities *De Lege Ferenda*

This chapter will first provide a general framework of the regulative options to address the issue of AWS *de lege ferenda* (8.1). Next, it will discuss scholarship that deals with the substantive challenges AWS create and possible solutions (8.2). We acknowledge that there are issues such as analogical interpretation and evolutionary interpretation in which form and substance are intrinsically connected. Those issues will be dealt with jointly with the substantive issues (Topic 8.2). Finally, this chapter discusses the risk of fragmentation of IHL if a specific regime on AWS is developed (8.3).

Despite being tempting to adopt a one-size-fits-all approach and applying directly existing laws created with a human behavior paradigm to AI, this path might present undesired outcomes (Abbott, 2020, p. 3; Garcia, personal communication, June 13, 2022).[1]

---

[1] In a similar perspective, Eugênio Garcia, in his personal capacity stated during an interview for the author's Ph.D:
"the idea of most countries is that you have new international norms to regulate this hypothesis of having weapons deciding on their own. This is something new. It is not possible to simply use the IHL that already exists, which serves as a basis. It is a new thing that has a revolutionary character. You take humans out of the decision-making process about the use of force – this in ultimate terms is the greatest risk of humans losing control over the use of force. Of course I'm also thinking ahead – this is a long-term risk, that you have "thinking" machines that will decide (...) It will be very difficult to control, unpredictability is a problem, you have no precedent about this, so you have to have at least some kind of regulation that could include a total ban from the beginning or some kind of limits on the use of this type of system, if one day it is built and used in this totally autonomous way. We are currently at a threshold where some drones are practically reaching this stage. There was the case in Libya of the STM Kargu-2 expert panel, then the company itself said it had human control." My translation from the original in Portuguese:
"a ideia da maioria dos países é você ter novas normas internacionais para regular esta hipótese de você ter armas decidindo por conta própria. Isto é algo novo. Não da para simplesmente utilizar o DIH que já existe, que serve como base. É uma coisa nova que tem um caráter revolucionário. Você tira o ser humano do processo decisório sobre o uso da força- isto em termos últimos é o maior risco dos humanos perderem o controle sobre o uso da força. Claro que também estou pensando lá na frente- isto é um risco de longo prazo, de você ter máquinas "pensantes" que

(*Contd.*)

8.1 Regulative Options: Overview

At the CCW GGE, States have argued for different regulative options as regards AWS, including a legally binding instrument, political declaration, code of conduct (Garcia, 2019, p. 4),[2] sharing of good practices, analogical interpretation, and evolutionary interpretation.

8.1.1 States' Positions

States either oppose a legally binding instrument, argue for a two-tier approach[3] without being explicit on a binding instrument, or favor a legally binding instrument, which might or might not encompass the two-tier approach.

Among the States that oppose a legally binding instrument are the US, Russia, the UK, Israel, and South Korea. The US claims that best practices change over time and that States should aim to exchange good practices and not codify them (US Delegation, 2018a).[4] The US considers it premature to negotiate a treaty on AWS and affirms that IHL is adequate to address the situation (Wareham, 2020). Russia has consistently opposed proposals to negotiate a legally binding instrument on such weapons or other measures. It says that "existing international law, including international humanitarian law, has some very important restrictions that fully cover weapons systems that have high degrees of autonomy" (Wareham, 2020). The UK argues for a compendium of good practices (Wareham, 2020; UK Delegation, 2020).[5] Israel claims that a legally binding instrument on AWS is a

---

vão decidir (...) Será muito difícil controlar, imprevisibilidade é um problema, você não tem nenhum precedente sobre isto, então você tem que ter pelo menos algum tipo de regulação que pode incluir a proibição total desde o início ou algum tipo de limites à utilização deste tipo de sistema, se um dia ele for construído e utilizado desta forma totalmente autônoma. Nós estamos no momento em um limiar em que alguns drones já estão praticamente chegando neste estágio. Teve o caso da Líbia do painel de peritos do STM Kargu-2, depois a própria empresa disse que tinha humano no controle" (Garcia, personal communication, June 13, 2022). This is a citation from the author's Ph.D. dissertation.

2 Garcia affirms that "A few outcomes could arise out of the GGE, which arguably can just continue deliberations and postpone any committal decision. The possibility of a political declaration has been flagged, to be issued by all states or some of them, or a code of conduct compiling rules and principles applicable to autonomous weapons. A mandate can also be given for states to negotiate in earnest, so that meetings in Geneva may be conducive to a legally-binding instrument to ensure meaningful human control overcritical functions in lethal autonomous weapons systems" (Garcia, 2019, p. 4).

3 The two-tier or two-track approach means debating which AWS are prima facie unacceptable and prohibited, and which AWS need regulation.

4 The US Delegation wrote that "(...) rather than seeking to codify best practices or set new international standards, States should seek to exchange practice and implement holistic, proactive review processes that, are guided by the fundamental principles of the law of war" (US Delegation, 2018a).

5 According to Wareham, "In 2019,UK while stating that the existing legal framework suffices to deal with 'new capabilities' that the country "(...) does not seek to predetermine the exact format of any GGE outputs relating to a normative and operational framework; as the delegation from the United States has pointed out, form must follow substance." The UK suggests a compendium of good practices" (Wareham, 2020).

(Contd.)

"radical path" (Acheson, 2021). South Korea "(...) is worried about regulation hampering civilian AI research and use, and has also highlighted the benefits of defensive autonomous weapons" (Slijper, Beck and Kayser, 2019).

France proposes a non-legally binding political declaration (Wareham, 2020 p. 23). In a paper submitted in 2022, Finland, France, Germany, the Netherlands, Norway, Spain, and Sweden expressed support for the two-tier approach but were not explicit regarding supporting a legally binding instrument.

> In the framework of the GGE, States should commit to: (1) outlaw fully autonomous lethal weapons systems operating completely outside human control and a responsible chain of command, as well as (2) regulate other lethal weapons systems featuring autonomy in order to ensure compliance with the rules and principles of international humanitarian law, by preserving human responsibility and accountability, ensuring appropriate human control and implementing risk mitigation measures (Finland, France, Germany, the Netherlands, Norway, Spain, and Sweden delegations, 2022).

In 2016 and 2018, China called for a ban on AWS. It later stated that the proposal was restricted to the deployment of AWS and not to production. Nonetheless, since 2018 China has not repeated the claim for a ban on AWS (Wareham, 2020, p. 14). Internally China's AI development plan calls for regulation and an accountability system (People's Republic of China, 2017).[6] China's posture is of strategic ambiguity (Kania, 2018). In 2022, China, like France and others, argued for the two-tier approach, meaning that AWS that are prima facie unacceptable shall be prohibited, and the other AWS need to be regulated. The goal is "to ensure relevant weapons systems are secure, reliable, manageable and in line with international humanitarian law and other applicable international law" (People's Republic of China Delegation, 2022).

The NAM defends the urgent need for a legally binding instrument and adopts the two-track approach, stating that the instrument must contain prohibitions

---

According to the UK Delegation, "A compendium of good practice mapped against a weapon lifecycle would provide a clear framework for the operationalization of the guiding principles by states. Providing actionable guidance for policy, technical, and military stakeholders could encourage the adoption of national regulations designed to strengthen respect for international law and offer guidance for how this could be achieved throughout the weapon lifecycle" (UK Delegation, 2020).

In 2022, the UK Delegation also claimed: "It is proposed that the GGE commission a document that sets out guidelines, advice and best practices on how states should approach the development and use of emerging technologies in the area of LAWS at each stage of its lifecycle" (UK Delegation, 2022).

6 The Chinese State Council New Generation Artificial Intelligence Development Plan states "Conduct research on legal issues such as civil and criminal responsibility confirmation, protection of privacy and property, and information security utilization related to AI applications. Establish a traceability and accountability system and clarify the main body of AI and related rights, obligations, and responsibilities" (People's Republic of China, 2017).

and regulations (Bolivarian Republic of Venezuela on behalf of NAM, 2022).[7] Delegations of Brazil (Brazilian Delegation, 2020b), Argentina, Costa Rica, Ecuador, Guatemala, Kazakhstan, Nigeria, Panama, Peru, the Philippines, Sierra Leone, the State of Palestine, Uruguay (Argentina and others, 2022a and 2022c), Chile, and Mexico (Chile and Mexico Delegations, 2022) favor a legally binding instrument. In this sense, Argentina, Costa Rica, Ecuador, Nigeria, Panama, the Philippines, Sierra Leone, and Uruguay proposed to the GGE on LAWS a Draft Protocol VI in 2022 (Argentina and others, 2022b).

8.1.2 Reflection on the Regulative Options

From the State perspectives presented, we claim that the form to address AWS accountability gaps fall into three categories: (1) a legally binding instrument; (2) "soft law" which embraces a compendium of good practices, principles, or codes of conduct, guidelines; and (3) political declarations. We recall that there is a spectrum in which at one end is hard law (obligations contained in legally binding instruments or customary international law), and on the other is a political declaration. In between, there is "soft law" (Guzman 2011, 173) which are "Guidelines of behavior, such as those provided by treaties not yet in force, resolutions of the United Nations, or international conferences, that are not binding in themselves but are more than mere statements of political aspiration" (Law and Martin, 2009). Significantly enough, such paths are not mutually exclusive (GGE CCW, 2019a). Interpretative solutions of the existing law, such as analogical interpretation and evolutionary interpretation, are also forms of addressing gaps and, at the same time, providing substance to it. Considering their hybrid nature, and that for them to work as a form possibility for AWS they also have to function as substance, they will be dealt with in part 8.2.

A legally binding instrument, such as an additional protocol, offers the advantage of providing a higher threshold of legal certainty. On the other hand, it bears two main risks. First, it is argued that it might freeze definitions and patterns that are constantly evolving in the field of AWS. Second, the most militarized States are unlikely to sign and ratify it.

Crootof addresses the first concern by suggesting that "(…) States negotiate a broad framework convention that can be augmented and expanded by specialized additional protocols, soft and interstitial law, and domestic law" (Crootof, 2016, p. 1397). She further argues that one of those complimentary protocols could contemplate international and domestic liability (Crootof, 2016, p. 1397).[8]

---

7 According to the document, the NAM States Parties and other States Parties to the CCW are of the view that given the substantive discussions in the GGE on LAWS, there is an urgent need to pursue a legally binding instrument under the Convention that will contain prohibitions and regulations for addressing the humanitarian and international security challenges posed by emerging technologies in the area of LAWS" (Bolivarian Republic of Venezuela on behalf of NAM, 2022).

8 Rebecca Crootof argues that "One of these additional protocols could outline an integrated international and domestic liability regime for autonomous weapon systems. At the very least, it should reiterate and clarify the relevance of the law of state responsibility. It could also clarify

(Contd.)

On the second concern, it is significant to highlight that so far, out of the P5+ Israel and South Korea, only China seems to be open to such a path. On the other hand, many States call for a legally binding instrument, such as those members of the non-aligned movement.

> It's possible that a subset of states would create a treaty outside of the UN CCW process that would ban AWS for those States, but that's unlikely to include leading military powers or leading robotics developers. I'm not sure that it changes the overall international trajectory other than in some intangible sense, perhaps contributing to some normative effect, further stigmatizing autonomous weapons, but I'm not sure how decisive of an effect that would have on actually slowing the development of the technology and its use (Scharre, personal communication, May 27, 2022).

It seems that the most militarized states such as the US, the UK, France, Russia, Israel, and South Korea would at the limit be open to a "soft law solution" or rather would prefer a political declaration. Soft law offers and political declarations are more flexible if technological and political situations change. They thus address State concerns that a specific regulation may not be desirable or adequate in the future. The disadvantage is that they offer less certainty, fewer or no sanctions, and are not as effective in deterring breaches. However, it is precisely because of those features that States are more likely to adhere to them (Guzman, 2011).

Despite the resistance of crucial actors, we claim that a legally binding instrument is the best path considering the risks AWS pose and the necessity to ensure responsibility, deter breaches, and have a coercive effect. We recall that the US has the 3009 Directive,[9] the UK has the Joint Doctrine 2/11, replaced by Joint Doctrine Note 30.2/2017, and France has the Opinion on the Integration of Autonomy into Lethal Weapon Systems (French Ministère des Armées, 2021). The development of domestic AWS directives and policies facilitates the further development of a global normative framework (Vilmer, 2021). Considering the consensus rule and the opposition by key military States on a legally binding instrument, it is likely that States at CCW GGE agree at least as a first step on a soft law instrument, from which the 11 Guiding Principles can be deemed a starting point. Soft law approaches could be a path toward a legally binding instrument (Garcia, personal communication, June 13, 2022).[10]

---

common definitions, describe overarching regulatory aims, and require member states to pass legislation creating domestic liability for both war crimes and war torts (which may entail waiving sovereign immunity)" (Crootof, 2016, p. 1397).

Hammond also raises the possibility of a legally binding instrument and worried about justiciability, suggests a clause of compulsory ICJ jurisdiction or other international tribunal. On the justiciability we recall that it is not a problem specific to legally binding instruments, but to international law in general and that the legally binding path offers a way out since States can agree on dispute resolution mechanisms such as ICJ's compulsory jurisdiction (Hammond, 2015, p. 656).

9 For comments on the 2023 US Directive 3009 see Appendix N.

10 In this sense Eugênio Garcia, in his personal capacity, during an interview for the Author's Ph.D. stated that:

(*Contd.*)

8.2 Substance: Scholarship Frameworks

We will briefly present the analogical interpretation, evolutionary interpretation (as mentioned before, both embrace regulative options and substance), tort law approach, and administrative law approach since they might shed light on our path towards, *de lege ferenda*, finding measures to enhance State responsibility. Scholars have presented some possible frameworks to address AWS and AI challenges at large, not explicitly regarding State responsibility, AWS, and the ARSIWA. It is relevant to note that many authors mention strict liability as a relevant ally. Considering the importance of strict liability to address many of the thirteen issues presented, it will be discussed in-depth in the next chapter.

8.2.1 Analogical Interpretation

Some scholars propose analogy as an essential resource to address AWS legal challenges[11] without analyzing it as regards AWS and the ARSIWA. Analogy means "a comparison between two objects, or systems of objects, that highlights respects in which they are thought to be similar" (Vöneky, 2008). In international law, analogical interpretation means to apply a rule that embraces a specific situation A to another situation B not addressed by the rule, but that has similarities with the situation. It is, for example, what has been done—to a large extent—concerning the application of international norms on physical spaces in the cyber domain (Schmitt, 2013).[12]

Analogical interpretation has 3 Steps. These are (1) abduction: "the moment when lawyers 'discover' the relevant similarity between a novel set of facts and

"The political reality is different. You have a group of states today led by the United States that have a proposal of a code of conduct. You will gather principles and that would be the most you can get in the current diplomatic environment. The GGE has 11 principles that were already adopted, but which are more abstract. This code of conduct, and also good practices, do not oblige the States. You can have a not-so-good practice and still claim that this is not a breach of any rule, as there is no such norm prohibiting it. Good practices and codes of conduct are non-binding, therefore weaker. You cannot apply sanctions. This is not meant to be the end point. These codes and principles are interesting, they help, but are insufficient from a legal point of view, including because you do not create a binding obligation." My translation from the original in Portuguese:

"Agora a realidade política é outra. Você tem um grupo de estados hoje liderado pelos Estados Unidos que têm uma proposta sobre a mesa de um código de conduta. Você vai reunir princípios e isto seria o máximo que você consegue no ambiente diplomático atual. Como o GGE tem 11 princípios que já adotou, mas que são mais abstratos. Este código de conduta, e também boas práticas, não obrigam os Estados. Você pode ter uma prática não tão boa e mesmo assim alegar que insto não é descumprimento de nenhuma norma, pois não existe tal norma proibindo. Boas práticas e códigos de conduta são não vinculantes, por isto mesmo mais débeis. Você não pode aplicar sanções. Este não é para ser o ponto de chegada. Estes códigos e princípios são interessantes, ajudam, mas são insuficientes do ponto jurídico inclusive porque você não cria uma obrigação vinculante" (Garcia, personal communication, June 13, 2022). This is a citation from the author's Ph.D. dissertation.

11 See for instance (Ford, 2017, p. 447). For further thinking on the theme (see Crootof, 2018).

12 See for example paragraph 14 of rule 30 Definition of Cyber Attack.

those covered by an existing rule"; (2) confirmation: reflect and adjust the rule chosen by abduction; (3) application of the analogical rule to the case (Bordin, 2018, p. 19).

As previously demonstrated through the thirteen issues of concern, there is a gap in State international responsibility regarding State breaches in the context of AWS. The difference between the situations addressed by the ARSIWA and AWS-related misdoing is that the ARSIWA is grounded on a human action paradigm, as previously discussed in Chapter 4, and thus covers State acts performed by humans. On the other hand, AWS-related misdoings are based on actions performed by autonomous systems that can select and engage targets without human input. Therefore, there might be a breach of the causal chain with human action. The similarity is that in both cases, State human-based actions and State AI actions, it is necessary to hold States responsible, and the ARSIWA provides a path to responsibility. Among the possible analogical frameworks for AWS, would be to base an analogy of AWS-related breaches on the rules applicable to combatants or child soldiers, for example.[13] The question here is whether the similarity is relevant enough to justify analogical interpretation (Bordin, 2018, p. 24).[14]

As regards State responsibility, if the set of rules chosen by abduction is the ARSIWA, they do not pass the filter of step 2, which is confirmation. An analogical interpretation seems too far a stretch to embrace AI misdoings in general, and AWS breaches specifically, as the whole system is grounded on a human action paradigm. While it could temporarily bridge a significant gap, it might be subjected to different interpretations by international and national courts and lead to significant uncertainty.

We recall that analogy bears a relevant role outside the field of adjudication, in the codification and development of new international law (Bordin, 2018, p. 47). Many articles of the ARSIWA themselves are the product of analogy with the Vienna Convention on the Law of Treaty (Bordin, 2018, pp. 36–37).[15]

Analogical interpretation also presents the advantage of addressing legal lacunes. Therefore, it develops an essential role in maintaining the systematicity of the legal system, and thus, well-grounded analogies are considered authoritative (Bordin, 2018, p. 27).[16] The question is "when is an analogy 'well-drawn'? Formal

---

13 To deepen in the issue of analogy in the context of AWS, see (Crootof, 2018).

14 Bordin explains: "Analogies start with an abduction which must be confirmed by analogy-warranting rationales that provide a normative explanation of why the lawyer drawing the analogy has effected 'an acceptable sorting of a range of particular items'" (Bordin, 2018, p. 24).

15 Bordin cites, for instance that Article 49(3) ARSIWA, was drafted based on Article 72(2) VCLT, and Article 42 ARSIWA was based on Article 60 VCLT (Bordin, 2018, pp. 36–37).

16 Bordin acknowledges "on the one hand, that international law does not constitute a legal system in the same way as domestic law, or that there is no such thing as an 'international rule of law'. On the other hand, international law deploys systemic reasoning in forms that are seldom discussed in the domestic context, such as in the codification and progressive development of the law. The case for the use of analogy in international law thus requires further elaboration" (Bordin, 2018, p. 27).

justice requires us to treat cases alike, but it does not tell us which cases are like and which are not" (Bordin, 2018, p. 24).

The disadvantages of analogical interpretation are that, differently from the deductive process, where the rule covers the situation, analogical interpretation provides a partial justification for applying a rule or set of rules to the case in question (Bordin, 2018, p. 20). Thus, objectors argue that it does not create valid law.

Analogy also bears the risk of leading to the application of the residual rule that what is not prohibited is allowed, a maxim known as the Lotus principle (PCIJ, 1927, p. 18).[17] The ICJ has used the Lotus principle to bridge gaps as regards arms. In the Military and Paramilitary Activities in Nicaragua, the ICJ affirmed that "in international law there are no rules other than such rules as may be accepted by the State concerned, by treaty or otherwise whereby the level of armaments of a sovereign State can be limited, and this principle is valid for all States without exception" (ICJ, 1986, par. 269).[18] Ten years later, in the Advisory Opinion regarding the use of nuclear weapons, the ICJ highlighted that the unlawfulness of using certain weapons is grounded on prohibition rather than on the lack of authorization (ICJ, 1996, p. 25),[19] despite recognizing the anti-Lotus function of the Martens Clause (ICJ, 1996, p. 35).[20]

---

[17] The PCIJ stated "International law governs relations between independent States. The rules of law binding upon States therefore emanate from their own free will as expressed in conventions or by usages generally accepted as expressing principles of law and established in order to regulate the relations between these co-existing independent communities or with a view to the achievement of common aims. Restrictions upon the independence of States cannot therefore be presumed"(PCIJ, 1927, p. 18). On this issue, see also (Bordin, 2018, p. 34).

[18] According to the ICJ, "The Court now turns to another factor which bears both upon domestic policy and foreign policy. This is the militarization of Nicaragua, which the United States deems excessive and such as to prove its aggressive intent and in which it finds another argument to justify its activities with regard to Nicaragua. It is irrelevant and inappropriate, in the Court's opinion to pass upon this allegation of the United States since in international law there are no rules other than such rules as may be accepted by the State concerned, by treaty or otherwise whereby the level of armaments of a sovereign State can be limited, and this principle is valid for all States without exception" (ICJ, 1986, par. 269).

[19] According to the ICJ, "Par. 52 The Court notes by way of introduction that international customary and treaty law does not contain any specific prescription authorizing the threat or use of nuclear weapons or any other weapon in general or in certain circumstances, in particular those of the exercise of legitimate self-defense. Nor, however, is there any principle or rule of international law which would make the legality of the threat or use of nuclear weapons or of any other weapons dependent on a specific authorization. State practice shows that the illegality of the use of certain weapons as such does not result from an absence of authorization but, on the contrary, is formulated in terms of prohibition" (ICJ, 1996, p. 25).

[20] According to the ICJ, "The Court would likewise refer, in relation to these principles, to the Martens Clause, which was first included in the Hague Convention II with Respect to the Laws and Customs of War on Land of 1899 and which has proved to be an effective means of addressing the rapid evolution of military technology. A modern version of that clause is to be found in Article 1, paragraph 2, of Additional Protocol 1 of 1977, which reads as follows: "In cases not covered by this Protocol or by other international agreements, civilians and combatants remain under the protection and authority of the principles of international law derived from established custom, from the principles of humanity and from the dictates of public conscience."

(*Contd.*)

Significantly enough, the Lotus principle is residual and not the way to bridge all gaps in the international legal system. "Whether the law is silent can only be ascertained by a thorough identification of existing rules and principles, of which the deployment of the techniques of legal reasoning – including argument by analogy – is an integral part." In short, the Lotus principle might only prevail if it is in accordance with the international legal system as a whole (Bordin, 2018, p. 35), including the Martens Clause.[21]

In the specific case of AWS, applying the Lotus principle without a specific regulation of AWS might mean broad permission for AWS and, thus, endanger international security. Even if the Martens Clause serves as an antigen to the Lotus principle, analogical interpretation could lead to divisive interpretations and picking and choosing.

In our perspective, analogy is not a suitable path as regards AWS for three main reasons: (1) it misrepresents unique features of AWS (Crootof, 2018, p. 83);[22] (2) it can limit the appropriate regulation of new technology since it restricts the potential envisioning of upcoming evolutions; and (3) it might be used as a means to shield decisions based on extra-legal grounds, which in the field of security might mean States picking and choosing the most convenient analogies. In this regard, we agree with Crootof's argument:

> Instead of borrowing from different legal regimes and relying on analogical reasoning to create a patchwork of regulations for autonomous weapon systems, we should instead acknowledge the fundamentally distinct nature of these new war fighters and create appropriate, supplemental law (Crootof, 2018, p. 83).

Whereas analogical interpretation can be a temporary and uncertain way to provide solutions in cases of gaps by avoiding non liquet, especially in the adjudication sphere, it might prove too uncertain. In our view, in the context of AWS, the analogy is best employed to refine reasoning about the suggestions

---

In conformity with the aforementioned principles, humanitarian law, at a very early stage, prohibited certain types of weapons either because of their indiscriminate effect on combatants and civilians or because of the unnecessary suffering caused to combatants, that is to say, a harm greater than that unavoidable to achieve legitimate military objectives. If an envisaged use of weapons would not meet the requirements of humanitarian law, a threat to engage in such use would also be contrary to that law" (ICJ, 1996, p. 35, par. 78).

21 For a deep discussion on AWS and the Martens Clause, see Docherty stating that "Existing treaties only regulate fully autonomous weapons in general terms, and thus an assessment of the weapons should take the Martens Clause into account. Because fully autonomous weapons raise concerns under both the principles of humanity and the dictates of public conscience, the Martens Clause points to the urgent need to adopt a specific international agreement on the emerging technology. To eliminate any uncertainty and comply with the elements of the Martens Clause, the new instrument should take the form of a preemptive ban on the development, production, and use of fully autonomous weapons" (Docherty 2018, p. 44).

22 In Crootof's words: "By misrepresenting legally salient traits and by limiting our ability to imagine future developments, analogies can sometimes impede our ability to appropriately regulate new technology" (Crootof, 2018, p. 84).

de lege ferenda in the progressive development of international law. A precise solution *de lege ferenda* is to provide the international community with specific rules as regards AWS.

In short, we believe analogy is not a tool to solve the challenges AWS manifest. The better path is acknowledging the unique traits of AWS and creating appropriate complementary laws (Crootof, 2018, p. 83).

8.2.2 Evolutionary Interpretation

Evolutionary interpretation is not proposed by the literature reviewed in the field of AWS. However, considering its significance in bridging gaps in evolving situations, it will be discussed here. Evolutionary interpretation is a process through which "the meaning of treaty terms may be liable to change over time, without the specific intervention of the parties to amend or modify the treaty terms" (Bjorge, 2014). It is discussed in the field of treaties, but we argue it is also applicable to interpret the ARSIWA due to its authoritative nature, as exposed in Chapter 3 of Part II.

As previously demonstrated through the thirteen issues of concern, there is a gap regarding the misdoings of State autonomous devices. The ARSIWA is grounded on a human action paradigm, as discussed in Chapter 4, and thus covers State acts performed by humans. One could question if the intention of the parties during discussions about the ARSIWA and its adoption by the General Assembly embraced autonomous devices misdoings, in other words, if they considered the AI revolution.

Evolutionary interpretation is not a new method of interpretation, and it is closely linked to intention and results from the "proper application of the usual means of interpretation, as means by which to establish the intention of the parties" (Bjorge, 2014). It builds upon the interpretation's traditional framework, which has as its object "to give effect to the intention of the parties as fully and fairly as possible" (Brierly and Clapham 2012, p. 349). The ILC report highlights that the interpretation of treaty terms is capable of evolving over time, and later agreements and practice, as stated in articles 31 and 32 of the Vienna Convention on the Law of Treaties (Vienna Convention on the Law of Treaties 1969), may be of use to define if the "presumed intention of the parties upon the conclusion of the treaty was to give a term used a meaning which is capable of evolving over time" (UN ILC, 2013).

Therefore, it has the advantage of making it possible to update agreements to societal changes without formal modification, helping to prevent them from becoming obsolete and ensuring that situations intended to be covered by the parties to an agreement do not remain unregulated.

Evolutionary interpretation has the disadvantage of encompassing a probable futuristic projection of parties' intentions. Therefore, except for a few prominent cases, it bears an inherent uncertainty and might be subject to different futuristic projections and lead to juridical uncertainty.

In our view, it cannot be claimed to be fully and fairly intended by the parties upon the ARSIWA's adoption by the General Assembly that States will accept autonomous devices misdoings. Autonomous device misdoings mean such a paradigm shift in the concept of human action that would probably lie outside the parties' intentions.

8.2.3 Tort Law Approach

Rebeca Crootof proposes a tort law approach (Crootof, 2016, p. 1387).[23] to address AWS challenges. War torts are grave breaches of IHL that trigger State responsibility (Crootof, 2016, p. 1386). Tort law focuses on redressing acts that cause damages and providing relief, regardless of whether they are wrongful acts.

The AWS individual accountability gap uncovers "(...) that, while there is an established (if not necessarily practically effective) means of holding individuals accountable for war crimes, the institutional processes of holding States accountable for their "war torts" are relatively undeveloped" (Crootof, 2016, p. 1353). The War torts regime provides a path to regulate important and intrinsically perilous activities (Crootof, 2016, p. 1353). It could clarify the law of State responsibility, incentivize the lawful behavior of States, deter them from using AWS that are likely to breach IHL and provide a path to compensation for victims of damages that do not originate from crimes (Crootof, 2016, p. 1398).

The strengths of the war torts regime proposed by Crootof would be to elucidate State responsibility in the context of armed conflict, especially the obligation to fully repair damages arising from violations of international law, by expressing which breaches are significantly severe to entail the obligation to make reparation (Crootof, 2016, pp. 1388–1389). Furthermore, "Public recognition of State fault and states' acceptance of responsibility would also entrench norms of lawful behavior" (Crootof, 2016, pp. 1388–1389). Hence, a war torts regime is expected to have a deterrent effect on States. Therefore, they would refrain from using means and methods of warfare that lead to severe violations of IHL. In addition, a war torts regime is likely to aid victims of international breaches by States to receive compensation, "which would not occur in the war crimes context. The fact that civilians are expected to shoulder the economic—to say nothing of the emotional—costs of the proportionality analysis is deeply troubling" (Crootof, 2016, pp. 1388–1389).

The tort law approach presents thoughtful insights, but also some limitations. In our perspective, the tort law approach does not address the specific challenges

[23] Crootof explains that "The various theories of tort law share a common assumption: that tort liability is grounded in a different kind of culpability than criminal law. While criminal law and tort law serve some of the same purposes—deterring undesirable actions through sanctions, holding those responsible for harm accountable, ingraining norms of conduct—the two legal regimes govern fundamentally different kinds of wrongs. Criminal law links legal culpability to moral culpability (...) Accordingly, some states refuse to create strict liability crimes entirely, on the grounds that it is incompatible with the nulla poena sine culpa principle—that there be no punishment without guilt" (Crootof, 2016, p. 1387).

autonomy poses to State responsibility raised through the 13 issues of concern, such as how to attribute responsibility to the State if there is no human action directly linked to the damage. How to allocate liability in the case of swarms composed by AWS belonging to different States?

Dickinson argues that what the proposal of "war torts" builds on current law is obscure since, under IHL, States are responsible for breaches, even if they are not crimes. She states that the ARSIWA offers a framework to hold States responsible for general breaches of international law (Dickinson, 2018, pp. 23–26).

Also countering Crootof's tort law approach are the ILC commentaries on the ARSIWA stating that in international law, it is unacceptable to distinguish, as in some domestic systems, among responsibility emerging ex contractu or ex delicto (UN ILC 2001b, p. 55). In this sense, the "Rainbow Warrior" arbitration is pertinent:

> The reason is that the general principles of International Law concerning State responsibility are equally applicable in the case of breach of treaty obligation, since in the international law field there is no distinction between contractual and tortious responsibility, so that any violation by a State of any obligation, of whatever origin, gives rise to State responsibility and consequently, to the duty of reparation (Report of International Arbitral Awards, 1990, p. 251).

Therefore, despite offering relevant insights into accountability challenges, the tort law approach, existent in some domestic systems, does not seem to address challenges generated by the paradigm shift in the concept of action that AI causes. Furthermore, it does not solve the specific challenges AWS pose to State responsibility.

8.2.4 Administrative Law Approach

Some authors propose using an administrative law approach to address AWS challenges. Some scholars propose a broader background of global administrative law, while others base their framework on domestic administrative law.

The administrative law approach analyzes the use of AWS as an "(…) exercise of state power against individuals through a computerized proxy" (Lieblich and Benvenisti, 2016, p. 246). While Lieblich and Benvenisti propose a broader background of global administrative law, Dickinson grounds her proposal on domestic administrative law.

According to Dickinson, criminal responsibility might not be possible in many AWS breaches, as it requires intent and causation. On the other hand, "Administrative accountability is flexible both in the process by which it unfolds and in the remedies available, offering the prospect of both individual sanctions as well as broader organizational reforms" (Dickinson, 2018, pp. 4–5).[24]

---

[24] According to Dickinson, "Perhaps international criminal law could be reformed to account for such issues. Or, in the alternative, greater emphasis on other forms of accountability, such

(*Contd.*)

Lieblich and Benvenisti affirm that administrative law lenses help to address the main challenges AWS raise. For them, modern warfare has to be understood not horizontally as an issue among sovereign States, but in the vertical framework as State power concerning individuals (Lieblich and Benvenisti, 2016, p. 263). Grounded on the legal obligation to exercise proper administrative discretion in targeting decisions, they argue that humans must make the final decisions (Lieblich and Benvenisti, 2016, p. 282).

This approach seems to disregard the unique feature of AWS, which is actions developed through AI, which entail autonomy and an inherent degree of unpredictability. Therefore, the concept of proper administrative discretion might not be applicable or needs to be adapted to this new paradigm of action, i.e., action perpetrated through autonomous devices.

Despite offering insights into accountability challenges, this approach does not address the root cause of the challenges AWS raise, which is the paradigm shift in the concept of action. This shift directly impacts, for instance, the attribution of conduct to the State, an issue not resolved by the administrative law perspective. The approach also disregards the inherent unpredictability of AWS while arguing for proper administrative discretion.

8.3 A Specific Regime of AWS Bears the Risk of Fragmentation of IHL?

Before thinking through the possibilities *de lege ferenda*, which will be done in the following chapters, this subheading analyzes if creating a special regime of State responsibility exclusively for a single type of weapon, AWS implies the risk of fragmentation of international law. It draws on the following questions: Is there a 'danger' of fragmentation? What would be the risks and rewards of creating this special regime? Do the benefits outweigh the risks? Is there a precedent in which a special liability regime was created for a specific type of weapon?

Fragmentation is international law's "institutional, procedural, and substantive diversification" (Peters, 2017, p. 678). Those who view fragmentation as a risk to international law argue that increasing subfields and international specific

---

as tort liability and state responsibility might be useful supplements. But largely absent from this debate is discussion of an alternative form of accountability that often gets overlooked or dismissed as inconsequential, one that we might term 'administrative accountability.' Such accountability includes multiple administrative procedures, inquiries, sanctions, and reforms that can be deployed within the military or the administrative state more broadly to respond to an incident in which a violation of IHL/LOAC may have occurred. This form of accountability may be particularly useful in the case of LAWS, because the restrictions of criminal law, such as the intent requirement for most crimes, may not apply in many circumstances. Administrative accountability is flexible both in the process by which it unfolds and in the remedies available, offering the prospect of both individual sanctions as well as broader organizational reforms" (Dickinson, 2018, pp. 4–5).

regulations could create friction and contradictions. States could face mutually excluding obligations and thus endanger the harmony, stability, cohesion, and consistency of international law as well as its comprehensive nature and legitimacy (Pellet, 2000, p. 26). The risk of divergence arises since there is no comprehensive internationally agreed regime to solve the conflict of norms.

Another view is that specialization is an opportunity to better accommodate specific concerns and that States are more prone to act according to international law with tailored rules (Hafner, 2004, p. 859).[25] Proliferation is also favorable since it demonstrates "the responsiveness of legal imagination to social change" (Koskenniemi and Leino, 2002, p. 575). Some authors argue that the fragmentation debate that started with trepidation around 2007 shifted, in the next decade, to praising and thinking through venues to coordinate and harmonize the specialized parts of international law (Peters, 2017, p. 674).

In the specific case of AWS, there is a risk that with a new treaty regulating AWS, for instance, competing norms would apply to different weapons during an armed conflict, or that different rules would apply to different parties depending on if they had or had not ratified a future treaty on AWS.

However, the specific features of AWS call for a specific regime. Since weapons can select and engage targets, there is such a paradigm shift in the concept of action that the benefits seem to outweigh the risks. Among the benefits is the increased chance that States comply with specific rules, since they are straightforward. States are more prone to adopt preventive measures to comply with them. Furthermore, a specific regime means the possibility of filling a responsibility gap by, for instance, assuring attribution to States and creating new opportunities for victims of breaches. Thus, an AWS regime is urgently needed to address its specificities.

To mitigate the risks of fragmentation, a treaty regulating AWS must state that it prevails if conflicting with other treaties, except if the other treaty ensures better protection of IHL and IHRL. In fact, "Systemic value could not be detached from the value of the system. If the law is unjust or unworkable, little virtue lies in applying it coherently" (Koskenniemi and Leino, 2002, p. 560).

Concrete examples of treaties that dealt with specific weapons are the Chemical Weapons Convention, the Treaty on the Non-Proliferation of Nuclear Weapons, and also the additional protocols to the CCW (Additional Protocol I – Non-Detectable Fragments, Protocol II – Prohibitions or Restrictions on the Use of Mines, Booby Traps and Other Devices, Protocol III – Prohibitions or Restrictions on the Use of Incendiary Weapons, and Protocol IV – Blinding Laser Weapons).

[25] Hafner states that "Specialization accommodates various needs and concerns of the States engaged in international law-making, and States perceive that their individual positions are better respected in these special regimes than in the global one. One may reasonably expect that, under such circumstances, States will be more induced to comply with these regulations and regimes" (Hafner, 2004, p. 859).

8.4 The Need for Adequate Paths to Address State Responsibility Challenges in Regarding AWS

Considering the thirteen issues of concern raised, that the existing scholarship does not address them efficiently, and also a discussion on the risks of fragmentation of IHL, we claim that it is necessary to find more suitable paths to ensure responsibility in the context of the novel paradigm AWS and AI present. Article 55 states that the ARSIWA are general rules and that the States can agree on specific rules (UN ILC, 2001c, art. 55). Special rules "(...) derogate from the general rules only to the extent stipulated for or else necessary for the implementation of the object and purpose of the special regime" (Kolb, 2017, pp. 72–73).

In the following chapters, we will offer some concrete suggestions, *de lege ferenda*, such as norms of a possible special regime and interpretative possibilities. The propositions are divided into three chapters (Chapters 9, 10, and 11). The first two chapters concern the main pillars to ensure responsibility for AWS breaches: a paradigm shift in attribution (Chapter 9) and strict liability as to the general framework (Chapter 10). Next, Chapter 11 presents a group of more specific observations/rules to address the issues presented on responsibility, which do not fall under attribution and strict liability, such as non-applicability of force majeure, a specific framework for human-machine interaction, weapons review, and due diligence.

CHAPTER

9

Issues that can be Addressed Through Tailored Attribution Rules

Issues that can be addressed through tailored attribution rules are:

- **Attribution Issue:** Considering that the ARSIWA is grounded on a human action paradigm, it is challenging to attribute responsibility to the State when the action or omission occurred in the context of AWS.
- **Due diligence Issue:** The breach of the duty of due diligence might be a ground to hold States responsible for their own acts and private acts, but the standard of due diligence, as regards AWS, needs to be established to enhance responsibility.

From all the challenges discussed, attribution seems to be the Achilles' heel for the responsibility of States regarding AWS breaches, and current rules are insufficient to ensure it. AI causes a metamorphosis in the concept of action that has so far been grounded on human action, as discussed in Chapter 4. This chapter discusses two possible solutions *de lege ferenda*: (1) a broader rule on attribution and (2) predefined schemes of attribution.

9.1 A Broader Attribution Rule

As discussed in Chapter 4, the ARSIWA is grounded on a human action paradigm and so far does not embrace AI actions.

To attribute the responsibility of acts in the context of AWS to States, a paradigm shift is necessary to include actions/omissions of AI within the scope of article 4 of the ARSIWA, even if they fall outside the sphere of foreseeability of a human actor.

In 2020, the US, aware of the attribution problem, proposed to the GGE the following conclusion "Under principles of State responsibility, every internationally wrongful act of a State, including such acts involving the use of emerging technologies in the area of LAWS, entails the international responsibility of that State" (US Delegation, 2020). The suggested draft conclusion proposed to include AWS within the internationally wrongful acts of the State that trigger responsibility. The US further suggested as a conclusion to the GGE that "A State remains responsible for all acts committed by persons forming part of its armed forces, including any such use of emerging technologies in the area of LAWS, in accordance with applicable international law" (US Delegation, 2020). This proposed text aims to ensure that acts of armed forces also include actions taken by AWS. The US proposal for the GGE seems to be precisely a push to a shift so that acts of AWS entail State international responsibility.

Following the US's suggestion, the GGE's Sep. 2021 Chair's paper entitled "Draft elements on possible consensus recommendations in relation to the clarification, consideration, and development of aspects of the normative and operational framework on emerging technologies in the area of lethal autonomous weapons systems" suggested that:

> "9. Under principles of State responsibility, any internationally wrongful act of a State or attributable to a State, including such acts involving the use of emerging technologies in the area of lethal autonomous weapons systems, entails the international responsibility of that State.
>
> 10. A State remains responsible for, inter alia, all acts committed by its organs including members of its armed forces, including any such use of emerging technologies in the area of lethal autonomous weapons systems, in accordance with applicable international law" (Chair of the 2021 GGE, 2021).

Corroborating this viewpoint, the 2022 CCW Chair's draft proposal was that:

> Every internationally wrongful act of a State, including such conduct involving weapon systems based on emerging technologies in the area of LAWS entails international responsibility of that State. The conduct of a State's organs such as its agents and all persons forming part of its armed forces, is attributable to that State, including any such acts and omissions involving the use of a weapons system based on emerging technologies in the area of LAWS, in accordance with applicable international law (Chair of the 2022 GGE, 2022).

A similar claim was made in the Principles and Good Practices on Emerging Technologies in the Area of Lethal Autonomous Weapons Systems (Australia and others, 2022),[1] and The Roadmap Towards New Protocol on Autonomous

---

[1] The document addressed attribution by stating that "The conduct of a State's organs such as its agents and all persons forming part of its armed forces, is attributable to the State. This includes any such acts and omissions involving the use of a weapons system based on emerging technologies in the area of LAWS, in accordance with applicable international law."

"18. State Responsibility. Under principles of State responsibility:

a. Every internationally wrongful act of a State, including such conduct involving the use of a weapons system based on emerging technologies in the area of LAWS, entails the international responsibility of that State.

b. The conduct of a State's organs such as its agents and all persons forming part of its armed forces, is attributable to the State. This includes any such acts and omissions involving the use

(*Contd.*)

Weapons Systems, concluding that acts and omissions involving the use of AWS are included within the framework of a State's organs being attributable to the State (Argentina and others, 2022a).[2]

However, until March 2023, despite recognizing State responsibility (GGE CCW, 2022b),[3] the GGE has not agreed on such a broader attribution provision.

Although not solving the attribution issue *de lege lata*, the provision on the excess of the authority of article 7 of the ARSIWA also points out that it is reasonable to include AWS actions within those attributable to the State. It foresees that a State is responsible even if a person or entity exceeds its authority or acts against instruction (UN, ILC, 2001c, art. 7). The ILC's commentaries refer to article 91 of the Additional Protocol I to the Geneva Conventions as an example of article 7 (UN ILC, 2001b, p. 46).[4] Nonetheless, considering that a party to the conflict "shall be responsible for all acts by persons forming part of its armed forces," the humanitarian rule is broader than the rule of article 7 (Sassòli, 2002, pp. 405–406). What is relevant is that neither article 7 of the ARSIWA nor Article 91 of the Additional Protocol I, so far, embrace AWS, as previously discussed. On the other hand, those instruments are the linchpin of the reasonability of the claim, *de lege ferenda*, to a broader attribution standard. AWS should not receive a more lenient treatment than human acts. If AWS conduct is inherently unpredictable, it is reasonable to attribute it to the State, considering that actions of humans that exceed authority or contravene instructions are imputable to the State.

In summary, in line with those documents, which expanded the scope of article 4 of the ARSIWA, by including AWS in the concept of State organs, and based on article 7 of the ARSIWA we claim that it is feasible and advisable *de lege ferenda* that States agree that AWS conduct is comprised within article 4 of the ARSIWA.

---

of a weapons system based on emerging technologies in the area of LAWS, in accordance with applicable international law" (Australia and others 2022).

2 The document states: "21. Reaffirm that the conduct of a state's organs such as its agents and all persons forming part of its armed forces, is attributable to that state. In accordance with IHL, IHRL, and ICL, this includes any such acts and omissions involving the use of AWS" (Argentina and others 2022a).

3 According to the report, "For the purposes of its work, the Group recognized that every internationally wrongful act of a State, including those potentially involving weapons systems based on emerging technologies in the area of LAWS entails international responsibility of that State, in accordance with international law. In addition, States must comply with international humanitarian law. Humans responsible for the planning and conducting of attacks must comply with international humanitarian law" (GGE CCW, 2022b).

4 ILC commentaries state that "4) The modern rule is now firmly established in this sense by international jurisprudence, State practice and the writings of jurists. It is confirmed, for example, in article 91 of the Protocol Additional to the Geneva Conventions of 12 August 1949, and relating to the protection of victims of international armed conflicts (Additional Protocol I), which provides that: 'A Party to the conflict … shall be responsible for all acts committed by persons forming part of its armed forces': this clearly covers acts committed contrary to orders or instructions. The commentary notes that article 91 was adopted by consensus and 'correspond[s] to the general principles of law on international responsibility'" (UN ILC, 2001b, p. 46).

This path addresses the Attribution Issue. However, this forum does not solve the Due Diligence Issue[5] regarding the situation of activity by private actors for which States are accountable due to the breach of due diligence.

Specific rules seem to be a relevant possibility to address attribution (UN ILC 2001c, art. 55) both as regards State AWS misdoings and State due diligence responsibility as regards private actors' misdoings in the context of AWS. Concrete examples of special attribution rules are article 139 of the UNCLOS (UNCLOS, 1982, art. 139) and the provisions of Space law (Treaty on Principles Governing the Activities of States in the Exploration and Use of Outer Space, including the Moon and Other Celestial Bodies, 1967, art. VI).[6] Considering that AWS are the riskiest of the possible venues of AI that States may use, we argue that special rules of attribution similar to those of UNCLOS and space law are a forum to ensure attribution.

Article 139 of UNCLOS provides that States are responsible for ensuring that activities at the common heritage of humankind area are developed in conformity with the convention. This responsibility embraces activities carried out by the State, "or State enterprises or natural or juridical persons which possess the nationality of States Parties or are effectively controlled by them or their nationals." It further states that "(...) damage caused by the failure of a State Party or international organization to carry out its responsibilities under this Part shall entail liability." It exempts States from liability in the case of damage by the above-mentioned non-State actors if the State has taken "all necessary and appropriate measures to secure effective compliance with the duties under the UNCLOS" (UNCLOS, 1982, art 139).

Article VI of the Treaty on Principles Governing the Activities of States in the Exploration and Use of Outer Space, including the Moon and Other Celestial Bodies, provides that States are responsible for activities in outer space by governmental organizations and by non-State actors. They also state that non-

---

5 The breach of the duty of due diligence might be a ground to hold States responsible for their own acts and private acts, but the standard of due diligence, as regards AWS, needs to be established to enhance responsibility.

6 In this regard Kolb states that "(...) there are also special rules of attribution under particular international law sources, mainly treaty regimes. Article 55 of the ASR on *lex specialis* applies also to attribution issues. Examples given are article 139 of the UNCLOS (1982), which provides that when a State sponsors certain private activities in seabed mining, it becomes responsible for certain acts of the private operator. In legal terms, this means that some acts of the latter are attributed to the former. A similar rule exists under outer space law: the State is responsible for all damages flowing from outer space activities under its jurisdiction, whether governmental or private" (Kolb, 2017, pp. 72–73).
In the same sense, Crawford, and others, affirm "In the first place, the lex specialis may comprise an entirely new set of secondary rules of attribution introduced by treaty and applicable as between the parties – a special rule of attribution ratione personae. One example arises in the context of Article 139 of the United Nations Convention on the Law of the Sea, which provides that a state may act as 'sponsor' of a private entity undertakings eabed mining: in so doing, the state assumes responsibility for certain acts of the private operator" (Crawford and others, 2010, p. 114).

State actors' activity requires State "authorizations and continuing supervision" by the appropriate State Party to the Treaty (Treaty on Principles Governing the Activities of States in the Exploration and Use of Outer Space, including the Moon and Other Celestial Bodies, 1967, art. VI). Considering the necessity of expressly embracing AWS action as attributable to the State, that State-private partnerships and private actors' activities are a challenge for attribution, as described in Chapters 4 and 7, and also that non-State actors might use AWS,[7] we suggest inspired by the 2022 Chair's draft proposal and the two treaties mentioned above, a broader attribution rule for AWS.

The proposed rule should state that activities carried out by State-owned AWS or AWS deployed by the State, State enterprises, State-private partnerships, and State sponsored activities are attributable to that State.

The rationale for the broader rules of attribution to embrace private actions is the inherent risk of allowing AWS and the State's duty to review weapons, discussed in Chapter 7.

In the case of both public and privately owned AWS, their development and use by State, non-State actors, or public-private partnerships requires State authorization, continuing supervision, and establishment of rules. Therefore, if States breach the due diligence duty of authorizing establishing rules and continuously supervising, responsibility is attributable to the State.

The due diligence framework is discussed in a *de lege ferenda* perspective in Chapter 11.7. In contrast to Article VI of the Treaty on Principles Governing the Activities of States in the Exploration and Use of Outer Space, including the Moon and Other Celestial Bodies, in the context of AWS, as regards private actors' activities, we believe a clarification of the duty of due diligence, namely authorization, continuing supervision, and establishment of rules the above-stated clarification of the context is necessary. Therefore, in the case of damage by non-State actors, if the State has taken its due diligence duty, it cannot be held responsible. We are not opting for direct attribution of private actors' AWS misdoings to the State since the situation on the ground in which an AWS is used, such as an armed conflict, is very different from the situation in outer space. It might happen, for instance, that a rebel non-State actor fights against the State. Therefore, it is reasonable to require specific due diligence from the State, but not direct attribution of private actor's activities.

Significantly enough, this broader attribution rule for AWS is a specific rule for a type of weapon, which creates an additional layer of juridical complexity to armed conflicts, as a rule of attribution would apply to traditional weapons, and a specific and broader rule applies to AWS. Therefore, there is a risk that certain attacks committed with traditional weapons are not attributable to the State, whereas AWS attacks are attributable, as the proposed AWS specific rule is broader. This risk of fragmentation exists, but its advantages seem to surpass the risks, as previously explained in Chapter 8.3, which discussed fragmentation,

7 For a deeper comprehension of non-state actors (see Chertof, 2018).

and also considering that attribution is one of the main challenges of AWS and that under the current regime, the chances of misdoings in the context of AWS being attributable to the State are much lower than internationally wrongful acts done with traditional weapons.

9.2 Predefined Schemes of Attribution

Another possibility to address the attribution challenges considering the human action paradigm of the ARSIWA is to predefine attribution in the context of AWS. This means shifting the focus to early-stage decisions, such as when humans activate the device, instead of focusing on critical functions and final decisions of selecting and engaging targets.

All militaries should have in place schemes of responsibility, meaning a predefinition of who is responsible for what, regardless of the weapons they use. However, the question is to what degree autonomy reshuffles schemes of responsibility (Bruun, personal communication, June 21, 2022),[8] and if it is possible to predefine schemes of attribution of responsibility in the context of AWS to determine who shall bear the burden if something malfunctions and to anticipate the focus on the last human touch point.

Considering that tracking which person caused a problem in the context of autonomous devices requires information on all the events and decisions taken and depending on the AI model used is not viable, a previously defined scheme of responsibility with each actor's responsibility and burdens and penalties would make the process of ascribing responsibility more manageable and add transparency (Prestes, personal communication, July 14, 2022).[9] Therefore, it is

8 This is a citation from the author's Ph.D. dissertation. During an interview for the author's Ph.D., Laura Bruun, in her personal capacity, stated: "A legal adviser from one State we talked to used the term of schemes of responsibility – a useful way to understand that you have to have some clear schemes in place – who does what, when and where – then you can also trace back (…) all militaries should have in place regardless of the weapons they use, but then we circle back to the question to what extent autonomy reshuffles this schemes" (Bruun, personal communication, June 21, 2022).

9 This is a citation from the author's Ph.D. dissertation. During an interview for the author's Ph.D., Edson Prestes, in his personal capacity, stated: "I think it's best to already have a well-defined accountability scheme. For me, it's much easier than trying to trace what caused that problem, because to trace, you need to have information about all the events, all the decisions made and, depending on the AI model you're working with. That's impossible, you can't. You can't tell if the problem is in the data, if the problem is in the lack of training. Then it becomes quite difficult. You can, of course, get this information but in a very superficial way. I believe that, if there is a model of accountability, previously defined, with the participation of each actor and with the possible penalties, it would make the process clearer and easier to deal with. It is worth emphasizing again that transparency and explainability, in the way we think, are beginning to make progress, but it is still very much in its infancy. Basic transparency and explainability can be achieved, as there are several levels, and can already be implemented and used on AWS." My translation from the original in Portuguese:
"Eu acho que o melhor é já ter um esquema de responsabilização já bem definido. Para mim, é bem mais fácil do que tentar fazer um rastreamento do que causou aquele problema, porque para fazer o rastreamento, tu precisa ter informações sobre todos os eventos, Todas as decisões

(Contd.)

a possibility to address the attribution challenge. The caveat is that schemes of responsibility would have to rely on the goodwill and different standards of States. Alternatively, it would have to be grounded on a juridical fiction of anticipating the critical moment to a preparatory one.

Within the framework of schemes of responsibility, we envision two possibilities of norms to address the Attribution and Due Diligence Issues:

Each State would be required to define its schemes of attribution of responsibility for its agents in the context of AWS, which would make possible the attribution of State responsibility. If they fail to do so, they will also be responsible.

Through juridical fiction, attribution in the context of AWS is shifted to early-stage decisions to deploy the weapon and not the critical functions of selecting and engaging targets.

The schemes should embrace not only State activities, but also who is in charge of due diligence as regards private actors' activities.

tomadas e, dependendo do modelo de IA que tu estejas trabalhando. Isso é impossível, tu não consegues. Não consegue saber se o problema está nos dados, se o problema está na falta de treinamento. Então se torna algo bem difícil. Tu consegues, é claro, obter esta informação mas de forma bem superficial. Eu acredito que, se existir um modelo de responsabilização, previamente definido, com a participação de cada ator e com as possíveis penalidades, tornaria o processo mais claro e fácil de se tratado. Vale ressaltar novamente que a transparência e explicabilidade, na forma como pensamos, estão começando a ter avanços, mas é algo muito ainda embrionário. Transparência e explicabilidade básicas podem ser obtidas, já que existem vários níveis, e já podem ser implementadas e usadas em AWS" (Prestes, personal communication, July 14, 2022).

CHAPTER

10

Issues that can be Addressed Through a Strict Liability Regime

Issues that can be addressed through tailored attribution rules are:

- **Intent and Breach Issue:** Rules of IHL and IHRL often require intent, which might hinder the very breach of an international obligation when AWS performed the conduct.
- ***Tempus Comissi Delicti* Issue:** Specific AWS features, such as opaqueness, add challenging roadblocks to assess the *tempus comissi delicti*.
- **Intent and Aiding Assistance Issue:** Considering that article 16 of the ARSIWA requires intentionality from State agents, responsibility for aiding assistance in the context of AWS might be challenging outside the framework of IHL.
- **Swarms and Allocation of Responsibility Issue:** Allocation of responsibility in the case of swarms composed by AWS from different States is uncertain under the ARSIWA, especially in the case of swarms that decide on a consensus basis or emergent coordination.
- **Programming and Authority Issue:** Even if international law evolves to consider AWS as State organs, it is unclear if AWS programmed by a sending State can be deemed under the authority of a receiving State.
- **Damage and Ascertaining the Wrongful Act Issue:** In the context of AWS and algorithmic decision-making, it is challenging to ascertain the wrongful act if there is no damage.

In this chapter, first, we will discuss the broad framework of strict liability (10.1), and next, (sections 10.2, 10.3, and 10.4) address the issues presented.

10.1 Strict Liability

In the framework of strict liability, damage gives rise to responsibility, and the issue of intentionality (intent, fault, negligence, recklessness) is removed from scrutiny. "Responsibility is triggered automatically whenever the risks inherent in unpredictable robotic behavior are realized" (Geiß, 2014, pp. 116–118).

Whereas under strict liability, a State responds for all injuries caused by its behaviors, under a negligence liability standard, responsibility only ensues if the injury results from failure to take adequate care (Crootof, 2016, p. 1394).

There is a trend at the national and international levels towards strict liability regimes regarding the development of dangerous activities (due to their nature or the means used to perform them) (Amoroso, 2020, p. 155). At the domestic level, States are discussing strict liability regimes regarding civil uses of autonomous technology (Geiß, 2014, pp. 116–118). At the international level, environmental international law has long-established strict liability for hazardous activities. The "polluter-pays" principle offers the basis for strict liability regimes (UNGA, 1992, art. 16; Convention on Civil Liability for Damage Resulting from Activities Dangerous to the Environment, 1993, preamble). In this regard, the International Convention on Civil Liability for Oil Pollution Damage (CLC, 1969) foresees that the ship owner is responsible for oil for pollution damage from the ship. It further states specific situations in which the owner is not responsible.[1] The Convention on International Liability for Damage Caused by Space Objects foresees strict liability in its article 20: "A launching State shall be absolutely liable to pay compensation for damage caused by its space object on the surface of the earth or to aircraft flight" (Convention on International Liability for Damage Caused by Space Objects, 1972, art. 2).

The ARSIWA does not generally foresee the threshold of responsibility, which depends mainly on the primary rules. Nonetheless, some provisions require a mental element for responsibility to ensue, such as article 16, which is not an outstanding primary norm, but a secondary rule that requires knowledge of the circumstances of the wrongful act, and the aim of facilitating the commission of the act, as will be further discussed in the next section.

---

1 For better comprehension, we transcribe part of the Convention "Article HI. 1. Except as provided in paragraphs 2 and 3 of this Article, the owner of a ship at the time of an incident, or where the incident consists of a series of occurrences at the time of the first such occurrence, shall be liable for any pollution damage caused by oil which has escaped or been discharged from the ship as a result of the incident.

2. No liability for pollution damage shall attach to the owner if he proves that the damage:

(a) resulted from an act of war, hostilities, civil war, insurrection, or a natural phenomenon of an exceptional, inevitable, and irresistible character, or

(b) was wholly caused by an act or omission done with intent to cause damage by a third party, or

(c) was wholly caused by the negligence or other wrongful act of any Government or other authority responsible for the maintenance of lights or other navigational aids in the exercise of that function.

3. If the owner proves that the pollution damage resulted wholly or partially either from an act or omission done with intent to cause damage by the person who suffered the damage or from the negligence of that person, the owner may be exonerated wholly or partially from his liability to such person.

4. No claim for compensation for pollution damage shall be made against the owner otherwise than in accordance with this Convention. No claim for pollution damage under this Convention or otherwise may be made against the servants or agents of the owner.

5. Nothing in this Convention shall prejudice any right of recourse of the owner against third parties" (CLC, 1969).

At the helm of IHL, Article 91 of Protocol I to the Geneva Conventions foresees that "A Party to the conflict which violates the provisions of the Conventions or of this Protocol shall, if the case demands, be liable to pay compensation. It shall be responsible for all acts committed by persons forming part of its armed forces" (Additional Protocol I, 1977, art. 91). Red Cross commentaries to article 91 claim that this article does not for the linchpin for strict liability (ICRC, 1987).[2] Moreover, as previously stated in Chapter 4, many humanitarian law provisions require intent. Thus, a strict liability regime is a possibility *de lege ferenda.*

Considering that activities involving AWS bear a degree of unpredictability, are ultra hazardous (Crootof, 2016, p. 1402), and are high risk, strict liability is the most appropriate regime to hold States accountable for AWS misdoings (Crootof, 2016, p. 1394). In the realm of State responsibility, differently from what occurs regarding individual criminal responsibility, "(...) an obligation to pay compensation could be validly grounded on ethical bases other than moral blameworthiness, such as duty to assume the risk for the harmful events caused by dangerous activities" (Amoroso, 2020, p. 253). As stated by Crootof:

> Unlike other weapons, autonomous weapon systems are capable of acting independently, breaking the causal chain between an individual's decision to deploy them and the target of these weapons' ultimate use of lethal force. And, unlike other robots, autonomous weapon systems are intended to kill people—they just are not supposed to kill the wrong people. The combination of these two factors strongly favor imposing strict liability (Crootof, 2016, p. 1402).

Identifying the lynch-pin of responsibility, only on the damage regardless of whether States were negligent or culpable, is relevant to attributing responsibility, especially in the case of technical failures, design defects, technical bugs, and accidents to States. Strict liability also incentivizes States to require excellent AWS development and use standards (Dutch AIV and CAVV, 2022, p. 39).

---

2 According to ICRC Commentary of 1987 on Article 91, "This responsibility covers 'all' acts committed by members of the armed forces of a Party to the conflict, and not only unlawful acts (or omissions conflicting with a duty to act) in the sense of the Conventions and the Protocol. We saw above that only acts which constitute violations (except for any case of 'force majeure' or fortuitous event that may occur) can give rise to compensation. However, it cannot be ruled out that the principle known as no-fault or strict liability would be taken into account, i.e., a concept of objective responsibility or liability which enters into play simply on the ground that the act or omission took place in the territory or under the jurisdiction of the State. This principle is recognized today in the field of environmental damage (irrespective even of the question whether there has been a breach of Article 35 [Link] – 'Basic rules,' paragraph 3), in nuclear matters, and in case of damage caused by spacecraft. In this sense it therefore seems possible that a Party to the conflict could be liable to pay compensation even in a case where no particular violation of the rules of the Conventions and the Protocol, or of another rule of the law of armed conflict, can be imputed to it. However, such liability could not be based on the present article" (ICRC, 1987, p. 1053).

Furthermore, considering that verifying the fault element is challenging, strict liability is an appropriate path (Amoroso, 2020, p. 155). It will be challenging to demonstrate that someone failed to exercise due care throughout the AWS cycle (manufacture, program, deployment) or link injury to a human failure (Crootof, 2016, p. 1395). If a negligence standard would be applicable, "(...), the responsible party could escape liability by conforming to the legal standard of care, even if there was no reason to engage in the activity in the first place" (Crootof, 2016, p. 1395). On the other hand, a strict liability standard allocates the burden of deploying AWS on States "(...) (subject to a contributory negligence defense)" (Crootof, 2016, p. 1396). State due diligence obligations are relevant but might be insufficient to ensure responsibility. In the context of strict liability, States might be held responsible even if they fulfilled their due diligence obligations.

A strict liability regime makes it easier to enforce responsibility, is more effective for victims, fosters diligent conduct by States, and impacts the frequency of use of AWS (Crootof, 2016, p. 1396).[3]

We claim that *de lege ferenda*, States agree on strict liability in the context of AWS and State responsibility (Dutch AIV and CAVV, 2022, p. 39).

10.1.1 Tailored Strict Liability Responsibility Regimes

In this section, we present some scholars' remarks that highlight AWS particularities and propose tailored regimes that include strict liability depending on certain circumstances. Next, we question if tailored regimes would be a better option than a pure strict State liability regime for AWS misdoings.

Focused on individual accountability of autonomous devices and not on State accountability for AWS, Abbott, based on his principle of AI neutrality, adds a caveat to strict liability – it should not be applied if AI is safer than humans are and humans are not subject to strict liability too (Abbott, 2020, p. 5). Despite recognizing the benefits of strict liability, he argues that if AI delivers better results and more safety than humans, for instance, in driving a car, it is not productive that human liability is based on fault and AI liability is strict liability. Abbott claims that AI responsibility should also be based on fault in such cases.

While Abbott's suggestion might be very relevant at the individual accountability level, we claim that it is not the best option for State liability. It would create much uncertainty or even a bar to responsibility to have to define in advance if human actions or AWS actions are safer to then be able to attribute it to the State or not.

Geiß bases his tailored accountability regime on a different perspective, not on AI neutrality like Abbott, but on the AWS specific feature of being designed to cause harm. For Geiß, considering that AWS are designed to cause harm, they

3 In an opposite perspective, Yavar Bathaee argues, with regard to AI in general, that considering AI unpredictability, a strict liability regime would not incentivize additional precaution (Bathaee, 2018).

are different from other strict liability regimes, such as the Outer Space Treaty and Space Liability Convention. Therefore, an absolute liability regime cannot be directly transposed to AWS (Geiß, 2016). In the context of armed conflicts, some damages are lawful, such as the destruction of a military objective, and the level of risk and unpredictability might depend on the kind of environment AWS are deployed in and on their interaction with the environment. He argues for a tailored regime that combines strict liability with other forms of liability.

> One could therefore envision a graduated liability regime whereby strict or presumed liability is imposed in certain scenarios or with respect to certain, fundamental rules, but not in all scenarios and not for the causation of damage generally, or in relation to the entire body of rules comprising the laws of armed conflict. Such a graduated liability regime that combines strict liability with other forms of liability may be best suited to respond to the different risks and uncertainties inherent in autonomous weapons systems that may be deployed in vastly different contexts (Geiß, 2016, pp. 116–118).

Significantly enough, this kind of tailored liability framework does not exist, and it is unlikely that States would agree on such a regime regarding AWS. It would bring about much uncertainty regarding which scenarios attract strict liability. Furthermore, the strict liability regime we propose does not ensue in every use of AWS. "If a state lawfully uses such a system against military targets, in full compliance with international humanitarian law, it is not responsible for the harm caused" (Dutch AIV and CAVV, 2022, p. 39).

In the context of personal responsibility and AI in general, and not based on the fact that AWS are designed to cause harm, Yavar Bathaee proposes a solution to address the black box and the gap of intent and causation in a sliding-scale model. In our view, Yavar's proposal offers some concreteness on a possible tailored regime suggested by Geiß. Yavar proposes to relax the requirements of intent and causation tests for liability when AI operates as a black box or without human control (autonomous) and maintains traditional intent and causation tests if AI is transparent or human-supervised (Bathaee, 2018, p. 936).[4] However, the

4 Bathaee affirms that "Putting the supervised and autonomous cases together, one can imagine four quadrants of liability. First, when there is both supervision of the AI and the AI is transparent, then the intent of the creator or user of the AI can be assessed through conventional means (i.e., fact-finding mechanisms such as depositions and subpoenas) as well as by examining the AI's function and effect. Second, when the AI is supervised but to some degree a black box, intent must be assessed based on whether the creator or user of the AI was justified in using the AI as he did – with limited insight into the AI's decision-making or effect. Third, if the AI is autonomous but supervised, the rule that should apply is the principal supervision rule from agency law. The question will be whether the creator or user of the AI exercised reasonable care in monitoring, constraining, designing, testing, or deploying the AI. Fourth, when the AI is both autonomous and unsupervised, the sole question will be whether it was reasonable to have deployed such AI at all. The answer may simply be no, which means that the creator or user of the AI would be liable for the AI's effects, even if he could not foresee them and did not intend them" (Bathaee, 2018, p. 936).

"relaxed" mode of intent and causation is not a strict liability but a negligence standard.

We argue that Geiß and Yavar Bathaee's models could inspire a tailored framework for individual civil accountability and corporate liability. Abbott's remarks are also essential to civil and corporate liability. Nonetheless, in the field of State liability, we argue for strict liability as to the general framework.

The reasons for not adopting such hybrid models are: States are in a privileged position to review the weapons, choose AWS, whether or not to deploy them, measure the possible impacts, and promote safe AI mechanisms. They are also benefiting from the military advantages of AWS. Moreover, strict liability may hinder technological development in the private sphere but not in the public. From a probationary perspective, such hybrid models might mean a roadblock to assuring State liability and redress to victims.

De lege ferenda, thus, we argue for a strict liability regime. It helps address accountability challenges that AWS pose due to their high risk, complexity, and inherent unpredictability. It would also incentivize States to take extra caution while deploying AWS and to "(…) take weapons reviews and steps towards risk minimization and harm prevention seriously, to program autonomous weapons systems 'conservatively,' and to implement strict product liability regimes domestically" (Geiß, 2016, pp. 116–118).

Even though strict liability eliminates an inquiry about culpability, strict liability does not always mean responsibility since a causal connection with the harm is still essential (Mann, 2019).[5] Suppose the AI algorithm changed meaningfully between the time of production/programming and the act/omission that led to the violation. It might not be possible to ascribe responsibility (Mann, 2019). Thus, it is necessary to ensure a relevant exception to strict liability when a break of the causal connection originated from a substantial alteration of the algorithm caused by external actors and was not within the possible foreseeability of the deploying State. Furthermore, if a State uses AWS lawfully, in total consonance with international law, against military targets, it is not responsible for the damage (Dutch AIV and CAVV, 2022, p. 39), as the damage is accepted by the law. In those cases, States bear the burden of proving that they are not responsible as they acted under international law. The contributory negligence defense also applies, meaning that if the victim contributed to the damage or injury, States might not be held responsible or only partially responsible.

From this general framework, in the following sections, we will provide some specific remarks regarding how strict liability can help address the abovementioned issues.

5 Talking about product liability and citing EU Directive 85/374/CEE of 1985, Mann emphasizes that a causal connection with the harm is still essential (Mann, 2019).

10.2 Strict Liability for AWS as a Primary Rule: Addressing the Intent and Breach Issue

The Intent and Breach Issue, as previously presented, was that rules of IHL and IHRL often require intent, which might hinder the very breach of an international obligation when conduct originated from AWS since they bear no mental element. Without a breach of an international obligation, it is impossible to ascribe State responsibility (see Chapter 4.2).

To attach State responsibility for breaches committed in the context of AWS, we argue for special rules that ensure strict liability for AWS misdoings. Therefore, even if primary rules require intent, through specific rules ensuring strict liability, it will be presumed in the case of AWS, making it unnecessary to demonstrate *mens rea*. Therefore, there will be a breach of international obligation, and State responsibility will be possible.

10.3 Strict Liability and Shared Responsibility: Addressing Intent and Aiding Assistance, Swarms and Allocation of Responsibility, and Programming and Authority Issues

Responsibility shared by States is a challenging issue under the ARSIWA.[6] AI adds a layer of difficulties to shared responsibility due to its inherent degree of unpredictability, and black boxes might be a burdensome roadblock to assessing the extent of responsibility of each State involved. As stated in the issues above, we identified three specific challenges: responsibility for aiding assistance, since article 16 requires intent; allocation of responsibility when swarms are composed of AWS from different States, and if AWS programmed by a sending State cannot be deemed under the authority of a receiving State. Each of them will be addressed in the next section. We also acknowledge that further studies might identify other issues regarding AWS and shared responsibility.

10.3.1 Strict Liability and Intent and Aiding Assistance Issue

The Intent and Aiding Assistance Issue stated that responsibility for aiding assistance might be burdensome outside the context of IHL, considering that article 16 of the ARSIWA requires intentionality, namely knowledge of the circumstances of the wrongful act and aims at facilitating the commission of the act, from State agents.

As discussed in Chapter 5.1, article 16 of the ARSIWA, requires a mental element, namely knowledge of the circumstances of the wrongful act, and aims at facilitating the commission of the act from the State agents. Thus, responsibility in

6 As in depth studied by SHARES Project (SHARES Project, n.d.).

the context of AWS misdoings might be challenging outside the context of IHL, where the obligation to respect and ensure respect for IHL applies. Also, despite the lower threshold, responsibility for aiding assistance might be challenging in the context of IHL, as it is necessary to demonstrate awareness of systematic violations.

> In armed conflict, under IHL's obligation to refrain from assisting in violations and "ensure respect" for humanitarian law, a State is responsible for providing weapons for a State that systematically commits violations of international humanitarian law with this kind of weapon, even if they could be used lawfully (Sossòli, 2002, p. 413).

De lege ferenda, we argue for strict liability as a special rule when the facilitation occurs in the context of AWS. This proposed special rule for AWS broadens the scope of article 16 of the ARSIWA and IHL to refrain from assisting in violations and "ensure respect." Therefore, attribution for aiding assistance in the context of AWS would depend only on demonstrating that a State aided assistance, without the necessity of intentionality. Thus, State responsibility ensues even without knowledge of the circumstances of the wrongful act, and the aim of facilitating the act's commission or the programmer State is aware that the State of destination systematically breaches humanitarian law.

10.3.2 Strict Liability and Swarms and Allocation of Responsibility Issue

The Swarms and Allocation of Responsibility Issue pointed out that allocation of responsibility in the case of swarms composed of AWS from different countries is also uncertain under the ARSIWA, especially in the case of swarms that decide on a consensus basis or emergent coordination.

In Chapter 5.3, we argued, *de lege lata*, on an interpretation of the principle of individual responsibility to allocate responsibility based on the kind of swarms. Nonetheless, we recognized that the issue might be subject to divergent interpretations and might be a source of controversy and accountability challenges. Determining and proving the extent of responsibility of each State in the case of swarms would be extremely hard.

A strict liability regime that expressly ensures shared responsibility to all the States that own devices that are members of the swarm, irrespective of the kind of swarm, would enhance responsibility *de lege ferenda*. Probing the degree of participation of each State would create a barrier to holding States accountable. If States decide to include a device which is their property as part of the swarm, an inherently risky activity, it is reasonable that they bear the burden of sharing strict liability.

10.3.3 Strict Liability and Programming and Authority Issue

The Programming and Authority Issue highlighted that even if international law evolves to consider AWS as State organs, it is unclear if AWS programmed by a sending State can be deemed to be under the authority of a receiving State.

We argued in Chapter 5.2 that AWS programmed by a sending State could not be deemed to be under the authority of a receiving State, in terms of article 6 of the ARSIWA, since the programming of AWS is a complex operation that might not be fully understood even to the programmers, and thus, will not likely be under exclusive direction and control of the receiving State.

De lege ferenda, a special rule assuring strict liability for receiving States that deploy AWS of other States, would clarify the above-stated situation, avoid a defense of lack of knowledge of programming, and blame-shifting. It would also create incentives for due diligence while choosing the device. If one State deploys AWS placed at its disposal by another State, it should be considered an act of the former. The receiving State benefits from having AWS at its disposal should bear the burden of responsibility if a breach occurs. Likewise, in the hypothesis that an AWS was programmed by a private actor from one State and sold to another State's government or private actor, the receiving State bears responsibility since it chose a private actor to do so and has to review and control weapons in its territory.

In sum, considering that responsibility for aiding assistance might be challenging in the context of AWS, article 16 of the ARSIWA requires *mens rea* from the State's agents, the challenges of responsibility in the context of multiple States' swarms, and that AWS activities are inherently risky, we argue for special rules that ensure strict liability as to the main framework for shared responsibility in the context of AWS.

10.4 Damage as a Trade-off for Strict Liability: Addressing Damage and Ascertaining the Wrongful Act Issue and *Tempus Comissi Delicti* Issue

The Damage and Ascertaining the Wrongful Act Issue emphasized that damage is not an essential element of the internationally wrongful act. However, from a probationary perspective, algorithmic decision-making adds an extra challenge in asserting the wrongful act if there is no damage. The *Tempus Comissi Delicti* issue highlighted that the specific features of AI add challenging roadblocks to assessing the *tempus comissi delicti*. We will deal with each of those topics from the perspective of requiring damage as a tradeoff for strict liability.

10.4.1 Damage as a Trade-off for Strict Liability: Damage and Ascertaining the Wrongful Act Issue

As discussed in Chapter 4.3, damage means loss or injury to a person or property. Damage is not an essential element of the internationally wrongful act and depends on the content of the specific primary rule. Considering their inherent unpredictability in the context of AWS, it might be very challenging to ascertain and prove a breach before damage occurs.

> "Empirically damage is a boosting factor for international responsibility, even if it is not, as it is known, a component of international responsibility. (…) Once you have damage, and this ascertained and objective, then probably States will be more likely to adopt new frameworks of responsibility such as strict liability" (Mauri, personal communication, July 5, 2022).

Supposing, *de lege ferenda*, a strict liability model is adopted; a possible tradeoff measure for this heightened accountability framework is to require damage for the breach to occur, or at least for strict liability to apply. First, it will be almost impossible in those situations to verify the occurrence of the breach if there is no damage. If the primary rules of AWS require damage, the probationary hardships presented in Chapter 4.3 would be solved. Furthermore, requiring damage increases the likelihood that States agree on a strict liability standard.

Note, for example, that the Convention on International Liability for Damage Caused by Space Objects successfully provides for strict liability and requires damage for the breach to occur. It states that the launching State is absolutely liable for "(…) damage caused by its space object on the surface of the earth or to aircraft flight" (Convention on International Liability for Damage Caused by Space Objects, 1972, art. 2).

Therefore, *de lege ferenda*, the proposal is that, specifically for AWS, States agree on a strict liability regime, requiring damage for the wrongful act.

10.4.2 Damage as a Tradeoff for Strict Liability: *Tempus Comissi Delicti* Issue

The *Tempus Comissi Delicti* Issue highlighted that AWS-specific features, such as opaqueness, add challenging roadblocks to assessing the *tempus comissi delicti*. We argued in Chapter 4.4 that the probationary field will be very complex or even unsolvable, especially considering the inherent unpredictability caused by the black box or interaction with the environment. If damage is required, many of those challenges would be addressed since the "act," as stated in article 14 of the ARSIWA, would be the immediate action that causes damage, and the challenges of setting the time of the breach would be mitigated. Regarding article 15 of the ARSIWA, which deals with composite acts, under the current framework, it will be hard to know when it suffices for the violation to occur and set the time of the breach. If damage is required, the composite acts suffice once damage occurs. Thus, many of the hardships of setting the time of violation in the AWS context will be addressed.

Therefore, we suggest *de lege ferenda,* a strict liability regime that would require damage, solving the challenges regarding *tempus comissi delicti.*

CHAPTER

11

Issues that can be Addressed Through Other Autonomous Weapons Systems Specific Rules

This chapter will present suggestions *de lege ferenda* to address the issues presented at the beginning of Part III, which may not be solved through attribution and strict liability norms.

11.1 Inapplicability of *Force Majeure* Solely Based on the Inherent Unpredictability of AWS: *Force Majeure* Issue

The *Force Majeure* Issue claimed that outside the context of peremptory norms, in general, the mere fact that violations were committed by AWS, which are inherently unpredictable, does not allow the defense argument of *force majeure* due to a State contribution to the breach. However, this could be interpreted differently. Furthermore, if the contribution was in good faith and unwilling, States can claim *force majeure* (Crootof, personal communication, May 26, 2022).[1] Another challenge is that the second hypothesis of inapplicability of *force majeure* requires an unequivocal risk assumption towards those to whom the obligation is due and will hardly be applicable.

De lege ferenda, we argue, based on the discussions of Chapter 6.1, and the inherent risks AWS present, that a specific norm that foresees the inapplicability

1 This is a citation from the author's Ph.D. dissertation. During an interview for the author's Ph.D., Rebecca Crootof, in her personal capacity, stated that: "States can still claim force majeure – that an act isn't wrongful due to an irresistible force or unforeseen event that makes it impossible for the State to perform its obligation. This defense only applies if the situation is not due to the acts of the State invoking the defense or if the State has not assumed the risk of the situation occurring. But those limitations often won't apply. First, States can claim this defense when the State has unwittingly contributed to the occurrence of the material impossibility, so long as its act was done in good faith and the act didn't make the bad event any more foreseeable. And second, you only get to say that the State has assumed the risk if the states unequivocally accept the risk of accidental harm. No State is going to say that it has done that. So, in many situations, States will still get to invoke force majeure" (Crootof, personal communication, May 26, 2022).

of *force majeure* due to the mere fact that AWS are inherently unpredictable is advisable. The rule should ideally state that *force majeure* does not apply to AWS breaches. Subsidiarily, such a norm could establish a presumption of knowledge of the contribution to the breach and risk assumption, being the burden of the State alleging *force majeure* to prove the contrary. Such a norm will prevent the *force majeure* issue from becoming a powerful defense argument that prevents or postpones State responsibility.

11.2 Addressing Assurance of Non-repetition Issue

The Assurance of Non-repetition Issue, as previously stated, was that AI and AWS also add additional challenges to assurance of non-repetition due to their inherent unpredictability.

As argued in section 6.2, a State could guarantee, for instance, adequate weapons review that a specific AWS or a specific algorithm would not be used again. However, it might be challenging to ensure that AWS would not incur the same breach outside those hypotheses. AWS bear an inherent degree of unpredictability, as explained in Part I. Thus, it is almost inviable to ensure non-repetition. In AI, the recent research area of AI safety aims to develop measures to lower the risk of AI devices. *De lege ferenda,* we suggest that preventive measures of AI safety, which means "the endeavor to ensure that AI is deployed in ways that do not harm humanity" (Feige, 2019), are required instead of assurances of non-repetition.

11.3 Addressing State Responsibility for not Using AWS Issue

The State Responsibility for not Using AWS Issue, as previously presented, was that not using AWS can *de lege lata* be regarded as a violation of international law if a State had the option of using AWS, and it was proven that its use would prevent a breach, as discussed in Chapter 7.2. However, considering the new ethical and legal challenges posed by autonomy, we argue *de lege ferenda* against holding States responsible for not using AWS.

Despite the obligation of using the more precise weapon, in our view, given the current state of technology, AWS cannot be considered a means of adding precision to an attack due to its features, such as autonomy, inherent unpredictability, and the opaqueness of AI decision-making. AWS shift the paradigm of action from humans to machines, expanding human reliance on complex systems (Wallach, 2015, p. 5),[2] and increasing risks of unexpected inaccuracy, as algorithmic decision-making follows a very different path than humans and might also be opaque to humans.

2 Wallach affirms that "Uncertainty in the development, progress and societal impact of an emerging technology is nothing new. However, our expanding reliance on complex systems whose risks we do not or cannot calculate is troubling" (Wallach, 2015, p. 5).

Removing or diminishing human decision-making and its unique capability (at least until now) of comprehending a broader context lowers precision when the broader context is considered. Furthermore, decisions produced by AI systems are based on the knowledge of the specialists and their programming or learned from a framework of previous or real-time collected data. A failure might mean large-scale damages as the system might operate in multiple machines simultaneously.

The prevention and precaution arguments are a fallacy in the context of AWS. Unlike other precise weapons, AWS might mean removing the most valuable asset on the battlefield, i.e., human reasoning, judgment, and comprehension of context. Furthermore, one of the main risks with autonomous technologies is that they are based on patterns and repeat those patterns until corrected. This exponential feature might lead to systemic and large-scale violations. Also, AWS bear an inherent degree of unpredictability, increasing the risk of unexpected IHL and IHRL violations. Therefore, even if proven that there is added precision in one specific situation, zooming out, there is a risk of exponential mass destruction and lack of comprehension of the broader context. It must also be recalled that some understand precision as a fallacy.[3]

De lege ferenda, it is necessary that a specific rule foresee no State responsibility and accountability for not using AWS, even if such use was likely to prevent breaches. The most valuable asset for human dignity is human reasoning and conscience (art. 1st DUDH). International law shall not incentivize that it be substituted by algorithmic decision-making, which would occur if States were held responsible for not using AWS. As discussed in Chapter 7.2, AWS cannot be deemed as more precise than other weapons, and their use does not add precaution or prevention. Furthermore, as with other autonomous technologies, possible AWS misfunctions might have an exponential effect and be deleterious until the bug is identified and addressed due to the pace of algorithmic decision-making, and the malfunction is replicated until corrected.

11.4 Addressing the Weapons Review Issue

The Weapons Review Issue stated that despite being a relevant source of State responsibility, the existing obligation to review weapons needs adaptations to better address AWS.

As discussed in Chapter 7.3, the existing obligation to review weapons, which applies to AWS just as it does to any other weapon, under article 36 of 1977 Additional Protocol I to the Geneva Conventions of 1949 or, at least its core aspect, which is a pre-deployment assessment of the legality of the weapon, under customary international law,[4] is an essential source of State responsibility. One

3 Note that some say that the most precise weapon is the landmine. It almost never fails and takes not only the injured human from the battlefield but also the two others who need to stop fighting to carry them.

4 Recall for instance that the US does not recognize the obligation to review weapons under customary international law, and does so on a policy basis.

of the significant challenges in weapons reviews is the inexistence of a general international framework, and each State reviews weapons in its own way.

De lege ferenda, the obligation to review AWS must be subjected to a general international framework that applies equally to all States to address the specific features of AWS, such as machine learning, and embrace AI safety mechanisms.

A periodic evaluation and improvement of AWS inputs and outputs are also necessary (Margulies, 2016, p. 19).[5] Weapons review embraces new and existing weapons that underwent significant changes after the review (Daoust, Coupland, and Ishoey, 2002, p. 352). Unless States agree on a ban on in-field learning, which is very unlikely, if the changes are significant, it is necessary to review the whole system. Even with minor changes, States might still have to review a changed algorithm. This specific AWS review process must answer questions such as, in the context of machine learning, when a new review is required. Shall States be obliged to review AWS periodically?

Furthermore, interpreting article 36 jointly with the obligation to respect and ensure respect for IHL,[6] States shall review the weapons they export (Daoust, Coupland and Ishoey, 2002, p. 352).[7]

De lege ferenda, among other issues, AWS review must include two specific issues that were touched upon by UN organs related to the human right to life and the human rights of persons with disabilities.

The impact on the human right to life must be a primary consideration in weapons review. Referring to article 36, the Committee on ICCPR exemplified the relationship between weapons review, autonomous weapons systems, and the right to life. "For example, the development of autonomous weapon systems lacking in human compassion and judgment raises difficult legal and ethical questions concerning the right to life, including questions relating to legal responsibility for their use" (UN ICCPRC, 2019).[8] Therefore, the Committee argues that AWS

---

5 Margulies states that "Approval of an AWS in the weapons review phase should be contingent on substantial ongoing human engagement with the weapons system. That human engagement has limits: A dynamic diligence approach will not require human ex ante authorization of AWS targeting decisions. However, this regime will require frequent, periodic assessment and, where necessary, adjustment of AWS inputs, outputs, and interface with human service members" (Margulies, 2016, p. 19).

6 In this regard, Article 1 is common to the four Geneva Conventions of 1949 and Additional Protocol I of 1977.

7 In a similar approach, Daoust, Coupland and Ishoey state "Although not specifically called for in Article 36, it would be desirable for States to examine also the legality of weapons to be exported. This would be in line with their obligation under Article 1 common to the four Geneva Conventions of 1949 and Additional Protocol I 'to respect and ensure respect' for these treaties" (Daoust, Coupland and Ishoey, 2002, p. 352).

8 For better comprehension, a longer part of the excerpt is transcribed "States parties engaged in the deployment, use, sale or purchase of existing weapons and in the study, development, acquisition or adoption of weapons, and means or methods of warfare, must always consider their impact on the right to life. For example, the development of autonomous weapon systems lacking in human compassion and judgment raises difficult legal and ethical questions concerning the right to life, including questions relating to legal responsibility for their use. The Committee is therefore of the view that such weapon systems should not be developed and put into operation, either in

(*Contd.*)

"should not be developed and put into operation, either in times of war or in times of peace, unless it has been established that their use conforms with article 6 and other relevant norms of international law" (UN ICCPRC, 2019).[9]

The human rights of persons with disabilities must also be a primary consideration, especially as to if the device can make a "nuanced determination as to whether an assistive device qualifies a person with disabilities as a threat" and also to assess if the device has the capability of assessing the reactions of persons with disabilities correctly (UNGA Human Rights Council, 2021).[10]

We offered some initial reflections since specifying the appropriate review process is beyond the object of the present research and can be, *per se*, the object of a book.

11.5 Time Frame Limitation and Human-Machine Interaction: Addressing (part of) *Tempus Comissi Delicti*, Weapon Review, and Human-Machine Interaction Issues

The *Tempus Comissi Delicti* Issue argued that specific AWS features, such as opaqueness, add challenging roadblocks to assess the *tempus comissi delicti*. The Human-Machine Interaction Issue is that International Law requires a human-machine interaction. Therefore, if a State fails to fulfill this requirement, it might be held responsible. However, the international community has to establish the threshold of this interaction and the kind of responsibility it gives

times of war or in times of peace, unless it has been established that their use conforms with article 6 and other relevant norms of international law" (UN ICCPRC, 2019).

9 For better comprehension, a longer part of the excerpt is transcribed "States parties engaged in the deployment, use, sale or purchase of existing weapons and in the study, development, acquisition or adoption of weapons, and means or methods of warfare, must always consider their impact on the right to life. For example, the development of autonomous weapon systems lacking in human compassion and judgment raises difficult legal and ethical questions concerning the right to life, including questions relating to legal responsibility for their use. The Committee is therefore of the view that such weapon systems should not be developed and put into operation, either in times of war or in times of peace, unless it has been established that their use conforms with article 6 and other relevant norms of international law" (UN ICCPR, 2019).

10 The Report of the Special Rapporteur on the rights of persons with disabilities stated: "54. At the same time, the use of artificial intelligence can have deleterious effects on persons with disabilities in situations of risk. For example, the deployment and use of fully autonomous weapons systems, like other artificial intelligence systems, raises concerns as to the ability of weaponry directed by artificial intelligence to discriminate between combatants and non-combatants, and make the nuanced determination as to whether an assistive device qualifies a person with disabilities as a threat. Further, the use of facial or emotion recognition technology at security checkpoints to assist in determining whether an individual may pose a threat lacks the same ability to correctly assess the reactions of persons with disabilities, owing to incomplete or biased data sets. To alleviate and address such concerns, persons with disabilities must be involved in the development, procurement and deployment of artificial intelligence technology as applied to situations of risk" (UNGA, Human Rights Council, 2022).

rise to. Furthermore, the Weapon Review Issue affirmed that despite being a relevant source of State responsibility, the existing obligation to review weapons needs adaptations to better address AWS.

While dealing with strict liability *de lege ferenda*, we suggested damage as a tradeoff for strict liability, which would help solve many of the issues regarding stating the tempus comissi delicti presented in Chapter 4.4.

Within the tempus comissi delicti issue, and the human-machine interaction issue, we suggest *de lege ferenda*, a time frame limitation to be established in the weapons review process. This limitation is likely to help assess the *tempus comissi delicti*, reduce causalities, corroborate the human-machine relation, and help ensure State responsibility.

To reduce the inherent unpredictability of AWS and enhance human-machine relations, every AWS should include a time limit. Margulies suggests, "The AWS should be programmed to move into a default hibernation mode after a relatively short, discrete period, which might be 24-96 hours" (Margulies, 2016, p. 22). Within this timeframe, "a human with remote access to the AWS's software could override the default and authorize continued operation for an additional increment of time" (Margulies, 2016, p. 22).

In a similar vein, the US DOD Directive 3009 requires that AWS can "complete engagements within a timeframe and geographic area, (…) consistent with commander and operator intentions. If unable to do so, the system will terminate engagements or obtain additional operator input before continuing the engagement" (US DoD, 2023).[11]

We argue for a timeframe limitation *de lege ferenda* as a significant hallmark in AWS regulation. What precisely this time limitation would be is outside the scope of the present research. States could either agree on a time frame or establish a time-frame umbrella under which each State would have to define the time frame limitation in the weapons review process.

11.6 Addressing the Human-Machine Interaction Issue

The Human-machine Interaction Issue, as previously stated, is that international law requires a human-machine interaction. Therefore, if a State fails to do so, it might be held responsible. However, the international community has to establish the threshold of this interaction and the kind of responsibility it gives rise to.

As discussed in Chapter 7.1 in the context of AWS, there exists an international consensus that some level of human involvement is necessary, but there is significant division about the degree of human involvement needed. Therefore, as a first step, we claimed to use the lower threshold of the appropriate level of human judgment. In this context, we proposed *de lege lata*, the interpretation that if a breach occurs, human agent responsibility is subjective,

[11] For comments on the 2023 US Directive 3009 see Appendix N.

and State responsibility is objective. This interpretation is based on the ARSIWA: "In the absence of any specific requirement of a mental element in terms of the primary obligation, it is only the act of a State that matters, independently of any intention" (ILC, 2001b, p. 36). However, if the primary rule requires intent, the human-machine interaction impacts the breach of an international obligation. In those cases, we have argued already *de lege lata* for a lower threshold, meaning that recklessness or negligence suffices for State responsibility to ensue, which is based on the due diligence obligation of States (that will be further discussed in the next section) and the inherent risk of activities involving AWS. *De lege ferenda*, we advocated for strict liability as argued in Chapter 10 of this Part III, which proposed to address the Intent Issue – "Rules of IHL and IHRL often require intent, which might hinder the very breach of an international obligation when AWS performed the conduct."

States are responsible for the acts of their AWS when there is a requirement for human judgment and State agents authorize the attack, even without authorizing every individual decision.

Furthermore, a legally binding instrument should assert that, at a minimum, States must implement mechanisms to ensure appropriate levels of human judgment. If States fail to do so, they can be held responsible for not complying with this obligation.

The challenge is to define what appropriate levels of human judgment (as a first step) mean and how to test if it suffices. *De lege ferenda*, ontological[12] patterns are an important alternative to provide certainty on what appropriate levels of human judgment mean and require.[13]

11.6.1 The Human-Machine Interaction Issue: Ontological Model

Ontological standards are models that formally name and define concepts, relations among them, and properties. Ontologies are helpful to standardize knowledge over one issue, based on stakeholder consensus, "defining, among other aspects, a standard knowledge structure in a domain, including common

[12] Ontological standards are models that represent, by formally naming, and defining concepts, relations, and properties, and the relation among them (IEEE, 2021a).

[13] In this sense, during an interview for the author's Ph.D., Edson Prestes, in his personal capacity, stated "That's where ontological patterns come in. They are important for giving more precision to real-world concepts and definitions. When we have a definition expressed in natural language, that definition [as in the case of meaningful human control], is very subject to ambiguity, and different interpretations. For example, what is significant in human control? Can we measure it? If so, at how many levels?" My translation from the original in Portuguese:
"É aí que entram os padrões ontológicos. Eles são importantes para dar mais precisão a conceitos e definições do mundo real. Quando a gente tem uma definição expressa em linguagem natural, essa definição, [como no caso de controle humano significativo] é muito sujeito a ambiguidade, e diferentes interpretações. Por exemplo, o que é significativo no controle humano? Podemos mensurá-lo? Se sim, em quantos níveis?" (Prestes, personal communication, July 14, 2022). This is a citation from the author's Ph.D. dissertation.

concepts, relationships, and attributes."[14] It also offers a formal framework for the standardization process, fostering a well-grounded standard. Furthermore, "the ontologies themselves, as formal artifacts, can be seen as products of the standardization process that can be used directly in data processing and automatic reasoning" (Prestes and others, 2021).

The model providing inspiration is the IEEE Standard 7007, which "is the first global ontological standard that contains the concepts, definitions, and axioms that are necessary to establish ethical methodologies for the design, development, and deployment of AI and robotics" (Prestes and others, 2021). It can be used as a foundation to develop public policies and elaborate AI systems. This model has as a goal:

> Establish a set of definitions and their relationships that will enable the development of robotics and automation systems in accordance with worldwide ethics and moral theories.
>
> Align the ethics and engineering communities to understand how to pragmatically design and implement these systems in unison.
>
> Develop a precise communication framework among global experts of different domains, including robotics, automation, and ethics (Prestes and others, 2021).

We claim that such a model is a backbone that would have to be specified for the context of AWS and human-machine interaction and also be grounded on ethical values and international law. It would make the human-machine relations and requirements explicit and could embrace both, a minimum threshold of appropriate levels of human judgment and an optimal threshold of meaningful human control.

> For example, an action, or a set of actions, is part of a plan. This plan was developed by a person A. When there is an ethical violation of a rule, there is a failure to satisfy a constraint that is linked to a given plan that was developed by one or more people. In that we are not looking directly at the artificial intelligence system, but at the structuring of how the information that is produced from the data collected by the system. This helps us to better understand what is happening or what happened in a given event (Prestes, personal communication, July 14, 2022).[15]

[14] Edson Prestes and others, "The First Global Ontological Standard for Ethically Driven Robotics and Automation Systems [Standards]" (2021) 28 IEEE Robotics and Automation Magazine 120. https://research-portal.uws.ac.uk/en/publications/the-first-global-ontological-standard-for-ethically-driven-roboti (Accessed 16 November 2022).

[15] This is a citation from the author's Ph.D. dissertation. During an interview for the author's Ph.D., Edson Prestes, in his personal capacity, stated "[ontological model] is path for more explainable AI. A starting point is to make system components explicit. For example, an action, or a set of actions, are part of a plan. This plan was developed by person A. When there is an ethical violation of a norm, there is a failure to satisfy a constraint that is attached to a given plan that was developed by one or more people. In what we are not looking directly at the artificial

(*Contd.*)

Prospectively, therefore, we claim for the development of a human-machine ontology that uses as a backbone the IEEE Standard 7007 (Prestes and others, 2013), which is the first global ontological standard that aims to establish methodologies for the development and ethical use of AI technologies. Ontological patterns are models that define concepts, the relationship between them, and their properties and are, therefore, an essential tool for standardizing knowledge on a topic (Prestes and others, 2021). It must also be based on UNESCO's recommendation on the Ethics of Artificial Intelligence (UNESCO, 2021, art. 35 and 36), and international law, especially IHL and IHRL. The proposed ontology should also be grounded on States, and other stakeholders' positions expressed both at the GGE and through policy documents, which would then be taken onto the CCW GGE floor for adjustments. This ontology would specify the minimal and the best human-machine relation in the context of AWS, based on predefined rules. After this process, it could also be transposed into a text and go through the process of becoming a legally binding instrument. This framework enhances State responsibility since it would materialize what those minimum requirements are; if States fail to do so, they can be held responsible. However, this ontology is outside the scope of the present book and has to be the object of further studies.

11.6.2 Human-Machine Interaction Issue and Explainable AI

Ontological standards are a path towards explainable or transparent AI, as they make explicit the human-machine relation and enhance responsibility. It is challenging to ascribe responsibility if what happens inside an AI system is an unknown.

> Transparency is a property that becomes increasingly essential as artificial intelligence systems interact with people in the real world. It will allow you to look at a situation and provide the system to determine which were the decisions taken by the system that led to a particular situation in order to, for instance, to determine the source of a failure. It is worth mentioning that there are different levels of transparency for different audiences. The reason is that the information an end user is interested in is different from what an auditor needs (Prestes, personal communication, July 14, 2022).[16]

---

intelligence system, but at the structuring of how the information is produced from the data collected by the system. It helps us better understand what is going on or what happened in a given event." My translation from the original in Portuguese:

"[modelo ontológico] é uma direção a ser seguida para IA mais explicável. Um ponto de partida é deixar os componentes do sistema explícitos. Por exemplo, uma ação, ou um conjunto de ações, são parte de um plano. Este plano foi desenvolvido por uma pessoa A. Quando existe uma violação ética de uma norma, existe uma falha em satisfazer uma restrição que está vinculada a um dado plano que foi elaborado por um a ou mais pessoas. No que não estamos olhando diretamente para o sistema de inteligência artificial, mas para a estruturação de como a informação que é produzida a partir dos dados coletados pelo sistema. Isso nos ajuda a compreender melhor o que está acontecendo ou o que aconteceu em um dado evento" (Prestes, personal communication, July 14, 2022).

16 This is a citation from the author's Ph.D. dissertation. During an interview for the author's (*Contd.*)

Transparency provides all the necessary components and data that help identify who is/are responsible in the case of failure. This means transparency provides information, but does not mean identifying a person responsible for the misdoings since, in the context of AI, technology is developed in teams, and also misdoings occur in a chain. For instance, a testing mistake might occur due to inaccurate data.[17] There are different levels of transparency, depending on various factors, such as the target audience (e.g., governments, programmers, operators). Prospectively, we claim that the ontological models proposed in the previous sections and that must be developed by further researchers ensure transparency as far as possible.[18]

11.7 Addressing the Due Diligence Issue

The Due Diligence Issue, as previously stated, is that the breach of the duty of due diligence might be a ground to hold States responsible for their own acts

---

Ph.D., Edson Prestes, in his personal capacity, stated "There are movements that try to say that transparency is something that is difficult to obtain, that it will reveal business secrets. However, in my view this movement goes in the direction of an evasion of responsibilities. Because, how can we assign responsibility in cases of failure if we don't know what really happened or is happening in the system?

Transparency is a property that becomes increasingly essential as artificial intelligence systems interact with people in the real world. It will allow you to look at the system, at a situation and provide subsidies to determine the origin of the problem. It is worth noting that there are different levels of transparency for different audiences; because the information an end user is interested in is different from what an auditor needs." My translation from the original in Portuguese:

"Existem movimentos que tentam dizer que transparência é algo que é difícil de ser obtida, que vai revelar segredos de negócio. Porém, na minha visão este movimento vai na direção de uma evasão de responsabilidades. Pois, como podemos atribuir responsabilidade em casos de falha se não sabemos o que realmente aconteceu ou está acontecendo no sistema?

A transparência é uma propriedade que torna-se cada vez mais essencial à medida que sistemas de inteligência artificial interagem com pessoas no mundo real. Ela vai permitir olhar para o sistema, para uma situação e fornecer subsídios para determinar a origem do problema. Vale ressaltar que existem diferentes níveis de transparência para diferentes públicos; pois a informação que um usuário final tem interesse é diferente daquela que um auditor necessita" (Prestes, personal communication, July 14, 2022).

17 In this sense, during an interview for the author's Ph.D., Edson Prestes, in his personal capacity, stated that "[even if the AI system is transparent] It will not identify a person, it will identify a group of people because you are always working in groups. And when a failure occurs, a chain failure occurs, in general. That is, the test failure may have happened because the data that was provided was not adequate data." My translation from the original in Portuguese:

"[mesmo que sistema de IA seja transparente] Não vai identificar uma pessoa, vai identificar um conjunto de pessoas porque tu sempre está trabalhando em grupos. E quando ocorre uma falha, ocorre uma falha em cadeia, em geral. Ou seja, a falha do teste pode ter acontecido porque os dados que foram fornecidos não eram dados adequados" (Prestes, personal communication, July 14, 2022). This is a citation from the author's Ph.D. dissertation.

18 Examples of transparency models (not directed to AWS) are the United Kingdom Centre for Data Ethics Innovation, Algorithmic Transparency Standard-Guidance for Public Sector Organizations (United Kingdom Centre for Data Ethics Innovation, 2023), and IEEE, Standard 7001 for Transparency of Autonomous Systems (IEEE, 2021b).

and private acts, but the standard of due diligence, as regards AWS, needs to be established to enhance responsibility.

As discussed in Chapter 7.4, the two conditions for responsibility to ensue under the breach of the duty of due diligence are: "(i) it had the means to prevent or to repress the unlawful act, and (ii) it knew or should have known about the risk of the violation" (Berkes, 2018, p. 445).

Regarding means to prevent, we claim that at the helm of LAWS, for acts committed by States themselves, the State deploying AWS, State-sponsored activities, and private actors' activities it seems reasonable to require the same level of due diligence for all States considering the ILA report on due diligence (French and Stephens, 2016, p. 18), the high risks AWS pose, the military advantages gained by States, and that they aimed at targeting an inherently risky activity, and also that most of the breaches occur in the framework of IHL.

The exception is that due diligence should vary depending on the parties' control over the territory or the role of non-State actors in the process (French and Stephens, 2016, p. 15). However, this could be interpreted differently; and we call for a specific statement *de lege ferenda.*

As regards the second requisite, State knowledge, we acknowledge the challenges since AWS are inherently unpredictable. However, States deploying or sponsoring or, for private actors' activities, authorizing AWS are required throughout the entire cycle of the AWS to test, evaluate, monitor, and anticipate possible damages. It is within this knowledge framework that responsibility for due diligence ensues. If this is not the case, there is no breach of the due diligence obligation. However, this could also be interpreted differently; therefore, *de lege ferenda* we suggest a rule recognizing a presumption of knowledge, shifting the burden of proof to States to demonstrate the lack of knowledge.

As observed in Chapter 7.4, due diligence is not specific in what is required (Bruun, personal communication, June 21, 2022),[19] and its contours in the context of AWS are yet to be defined by States, scholars, and courts. Therefore, we claim the necessity of establishing a minimum due diligence framework regarding AWS. The guiding question is: what minimum threshold of due diligence obligation actions/omissions in the context of AWS is required? Within this framework, we offer some suggestions without the aim of fully addressing the issue, which has to be the object of further studies.

To meet their due diligence obligations concerning AWS, States will have to enact domestic legislative and administrative measures to prevent breaches and damage, including, among other measures:

[19] This is a citation from the author's Ph.D. dissertation. During an interview for the author's Ph.D., Laura Bruun, in her personal capacity, stated: "State's responsibility applies not only for acts, but for omissions, and I think that it is all the things States are not going to do that will harm civilians in relation to AWS. All the due diligence obligations and precautionary measures that States have to take (In our work we talk about, just to give some examples, legal review, legal advice, training obligations, and instructions). All these obligations that States must implement to ensure that in the end no one is harmed. All these obligations are not really specific in terms of what they require or when a legal review is sufficient" (Bruun, personal communication, June 21, 2022).

- Require impact assessments and permit procedures for all activities involving AWS.[20]
- Establish monitoring mechanisms and monitor the implementation of the activities involving AWS as long as they continue.[21]
- Establish certification programs.[22]
- Conduct weapons review (as previously discussed in Chapter 7.3 of Part II and 11.4 of Part III).
- Take actions "to spare the civilian population, civilians and civilian objects. All feasible precautions must be taken to avoid, and in any event to minimize, incidental loss of civilian life, injury to civilians, and damage to civilian objects" (UK Delegation, 2022).
- Take precaution in attack and must "cancel or suspend an attack if it becomes apparent that the target is not a military objective or is subject to special protection, or that the attack may be expected to violate the rule of proportionality, as required by the rules on precautions in attack" (Davison, 2017).

Those initial due diligence guidelines must be the object of further work.

In sum, *de lege ferenda*

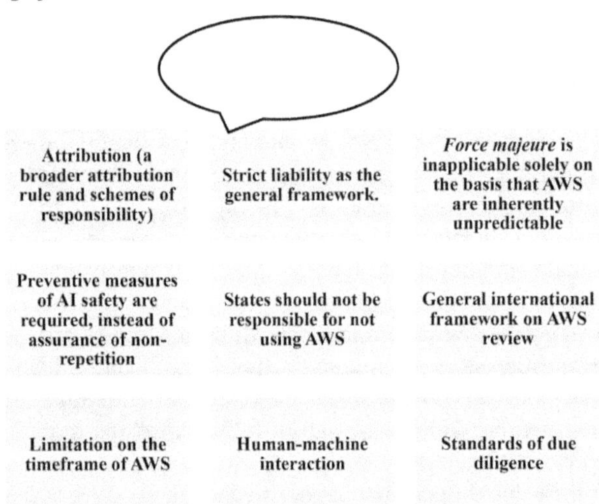

20 Inspired by the ILA report on the environmental law framework of due diligence (French and Stephens, 2014, p. 28) and ICJ case law (ICJ, 2015, p. 665, par. 153).

21 Inspired by the ILA report on the environmental law framework of due diligence (French and Stephens, 2014, p. 28) and ICJ case law (ICJ, 2015, p. 665, par. 153).

22 See, for example, Edson Prestes and others, stating that "The IEEE Ethics Certification Program for Autonomous and Intelligent Systems [6] is a world first in setting standards for the ethical certification of products, services, and systems deploying AI and robotics in the public and private sectors. Certification is essential to guarantee that these technologies operate as expected when they are interacting with human and nonhuman agents" (Edson Prestes and others, 2021, p. 121).

Conclusion

Part I provided a foundation for developing the research, presenting a concept of AWS, its inherent unpredictability, and the challenges to accountability. It explained that, so far, State responsibility for Autonomous Weapons Systems (AWS) breaches has been relatively ignored.

Part II consisted of an analysis of the ARSIWA with regard to AWS. It demonstrated that the current regime on State responsibility is insufficient to address the challenges posed by AWS.

Part III recalled the discussions of Part II and clustered thirteen issues of concern into three possible paths *de lege ferenda* to enhance State responsibility and accountability regarding AWS misdoings, which were schematically represented in the figure below.

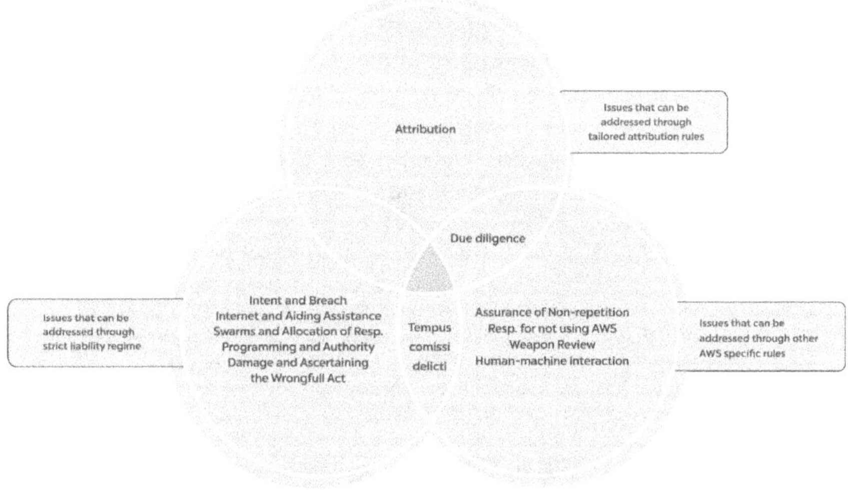

The two main paths suggested were: adjustments regarding attribution that consider the new paradigm of action AWS poses (a broader attribution rule and schemes of responsibility), and strict liability as the general framework for AWS misdoings. Requiring damage was also suggested as a trade-off for strict liability. Furthermore, concerning the issues not addressed by a shift in attribution and strict liability, seven specific suggestions were made *de lege ferenda*: force majeure is inapplicable solely on the basis that AWS are inherently unpredictable (11.1); preventive measures of AI safety are required, instead of assurance of non-repetition (11.2); States should not be held responsible for not using AWS even if it is demonstrated that their use would reduce causalities (11.3); the obligation to review AWS requires a general international framework, that addresses the specific features of AWS and embraces AI safety mechanisms (11.4); some limitations on the timeframe of AWS would enhance responsibility (11.5), the need to address the human-machine interaction to strengthen responsibility (11.6); and the necessity of establishing standards of due diligence (11.7).

In conclusion, Part II confirmed that the current international responsibility regime, as stated by the ARSIWA, is insufficient to address the challenges posed by AWS, and Part III provided some suggestions *de lege ferenda*.

Issues raised regarding AWS demonstrate a broader challenge that AI and autonomous devices pose to the international responsibility of States.[1] The military domain is not siloed, and the challenges of IA are often also relevant in other fields (Prestes, personal communication, July 14, 2022).[2]

Except for weapons review, all of the issues presented can be either directly applicable or adapted to other autonomous devices. The ARSIWA, as they currently stand, are not capable of addressing the challenges posed by either AWS or autonomous devices in general. Concerning AWS, a specific regime is necessary to address its misdoings. Concerning autonomous devices in general, it is necessary to have a broad discussion on autonomous devices in the process of turning the ARSIWA into a treaty. Alternatively, a special instrument can be carved into States' international responsibility in the context of autonomous devices. This book opens this discussion that has to be addressed by further scholarship.

As for the proposed paths *de lege ferenda*, the shift in the paradigm of action and a wider attribution framework is a cornerstone to ensure State responsibility, both for autonomous devices in general and for AWS. Those devices are making

---

1 As noted by Daniele Amoroso "(…)The legal problems raised by autonomy in weapons systems provide a uniquely representative sample, of the (potentially) disruptive impact of new technologies on norms and principles of international law". (Amoroso, 2020, p. 3).

2 This is a citation from the author´s Ph.D. dissertation. During an interview for the author´s Ph.D., Edson Prestes, in his personal capacity, stated that: "When we talk about AWS most want to discuss only topics associated with the military area. That is, it is as if the military area were self-contained, as if AWS and the problems associated with them happened only in that domain. And this is a very serious mistake because the problems that happen there happen in other fields. You have to look at the problems that are happening in other areas, and how they are being addressed, so that they are not replicated in the military domain." My translation from the original in Portuguese:"Quando falamos sobre AWS a maioria quer discutir apenas temas associados à área militar. Ou seja, é como se a área militar ela fosse autocontida, como se AWS e os problemas associados a elas acontecessem apenas naquele domínio. E isto é um erro muito grave porque os problemas que acontecem ali acontecem em outros campos. É preciso olhar os problemas que estão acontecendo em outras áreas, e como estão sendo tratados, para que eles não sejam replicados no domínio militar."(Prestes, personal communication, July 14, 2022)

decisions and performing activities that humans only developed in the recent past, and international law must evolve to address this venue of action. Strict liability as the main framework also seems to be the pillar to ensure responsibility for AWS misdoings, considering its ultrahazardous nature and that States who take advantage of them must bear the burden of a heightened liability standard. Moreover, it creates incentives for more diligent deployment and to implement safeguards. As for other autonomous devices, strict liability might be a possible path, but in-depth analyses need to be conducted in further works not to hinder technological development. Regarding the seven specific proposals, the following six can also apply to other autonomous devices: *force majeure* is inapplicable solely on the basis that autonomous devices are inherently unpredictable; that preventive measures of AI safety are required instead of assurances of non-repetition; States cannot be held accountable for not using autonomous devices; some sort of limitation on the timeframe of AWS is necessary to ensure responsibility; it is necessary to address the human-machine interaction by developing, for example, frameworks of ontologies; and it is necessary to agree on standards of due diligence.

AWS are, therefore, a case study of challenges posed by autonomous devices to the international responsibility of States, though the feature of dealing with visceral issues such as life and death often highlights the challenges of AI and autonomy. They bring to the surface old existing problems regarding State responsibility but also present new challenges generated primarily by the new paradigm of action of autonomous devices. We believe the present book has contributed to raising awareness of the State's responsibility gaps and opening the doors for discussions on *de lege ferenda* possibilities. Hopefully, the future implications of the present research will trigger discussion and action toward assuring State responsibility in the context of autonomous devices. Many issues require further research and development, such as guidelines on what due diligence requires in the context of AWS, what would be an adequate framework for weapons review as regards AWS, how we can draw an ontology that defines a minimum human-machine threshold for State responsibility, how to best define a time-frame limitation as regards AWS. The thematic calls for fostering interdisciplinarity as the legal and technical discussions often occur in silos and need closer interaction to address complex problems posed by AWS and other autonomous devices.

Appendix A

AWS Timeline[1]

2012	DoD Directive 3000.09, "Autonomy in Weapon Systems," November 21, 2012
2013	"Following a report by UN Special Rapporteur on extrajudicial, summary, or arbitrary executions Christof Heyns, the CCW Meeting of High Contracting Parties (HCP) decided that the Chairperson will convene in 2014 an Informal Meeting of Experts to discuss the questions related to emerging technologies in lethal autonomous weapons systems (LAWS)."
2014	The first Informal Meeting of Experts is held in accordance with the decision of the 2013 CCW Meeting of HCPs. *Chair: Ambassador Simon-Michel of France*
2015	The second Informal Meeting of Experts is held. *Chair: Ambassador Michael Biontino of Germany* Human Rights Watch issues the report Mind the Gap: The Lack of Accountability for Killer Robots.
2016	"The third Informal Meeting of Experts is held. *Chair: Ambassador Michael Biontino of Germany*"
2016	"At the CCW Fifth Review Conference, HCPs decide to establish an open-ended Group of Governmental Experts on Emerging Technologies in the Area of Lethal Autonomous Weapons Systems. *Mandate: to build on the work of the previous meetings of experts and to explore and agree on possible recommendations on options related to emerging technologies in the area of LAWS, in the context of the objectives and purposes of the Convention.*"
2017	CCW GGE Started to have formal meetings in 2017. "The Group of Governmental Experts meets for five days under the mandate agreed upon at the Fifth Review Conference. *Chair: Ambassador Amandeep Singh Gill of India*"
2018	"The Group of Governmental Experts meets for 10 days and affirms 10 guiding principles. *Chair: Ambassador Amandeep Singh Gill of India*"
2019	"The Group of Governmental Experts meets for seven days and identifies one additional guiding principle. The Group recommends the CCW HCPs endorse the 11 guiding principles at the annual Meeting. *Chair: Mr. Ljupco Jivan Gjorginski of North Macedonia*"
2019	"The CCW Meeting of the High Contracting Parties adopts 11 guiding principles per the recommendation of the 2019 GGE."

1 This is an edited version based on the UNODA's "Timeline of LAWS in the CCW." Retrieved from https://disarmament.unoda.org/timeline-of-laws-in-the-ccw/

2021	"The Group of Governmental Experts meets for 20 days and prepares the work of the CCW Sixth Review Conference, held in December 2021. No substantive report is adopted. *Chair: Ambassador Marc Pectseen de Buytswerve of Belgium*"
2021	"At the Sixth Review Conference, CCW HCPs, *inter alia*, affirm that international humanitarian law continues to apply fully to all weapons systems, including the potential development and use of LAWS, and that a lethal autonomous weapon must not be used if it is of a nature to cause superfluous injury or unnecessary suffering, or if it is inherently indiscriminate, or is otherwise incapable of being used in accordance with international humanitarian law. The Review Conference decides on the new mandate to: *'consider proposals and elaborate, by consensus, possible measures (...)'*".
2022	"The Group of Governmental Experts meets for ten days and decides to continue its work by intensifying the consideration of proposals. *Chair: Ambassador Flávio S. Damico of Brazil*" The groups report contains a Statement on State responsibility: "For the purposes of its work, the Group recognized that every internationally wrongful act of a State, including those potentially involving weapons systems based on emerging technologies in LAWS entails international responsibility of that State, in accordance with international law. In addition, States must comply with international humanitarian law. Humans responsible for the planning and conducting of attacks must comply with international humanitarian law."
2023	"The Group of Governmental Experts meets for 10 days and, inter alia, converges on the notion that weapons systems based on emerging technologies in the area of LAWS must not be used if they are incapable of being used in compliance with IHL. *Chair: Ambassador Flávio S. Damico of Brazil*" DoD Directive 3000.09, "Autonomy in Weapon Systems," January, 25, 2023 – Reissues and Cancels: DoD Directive 3000.09, "Autonomy in Weapon Systems," November 21, 2012.

Appendix B

Excerpts from the Report of the Special Rapporteur on Extrajudicial, Summary or Arbitrary Executions[1] on AWS

"Summary

Lethal autonomous robotics (LARs) are weapon systems that, once activated, can select, and engage targets without further human intervention. They raise far-reaching concerns about the protection of life during war and peace. This includes the question of the extent to which they can be programmed to comply with the requirements of international humanitarian law and the standards protecting life under international human rights law. Beyond this, their deployment may be unacceptable because no adequate system of legal accountability can be devised, and because robots should not have the power of life and death over human beings. The Special Rapporteur recommends that States establish national moratoria on aspects of LARs and calls for the establishment of a high level panel on LARs to articulate a policy for the international community on the issue."

(...)

"IV. Conclusions

109. There is clearly a strong case for approaching the possible introduction of LARs with great caution. If used, they could have far-reaching effects on societal values, including fundamentally on the protection and the value of life and on international stability and security. While it is not clear at present how LARs could be capable of satisfying IHL and IHRL requirements in many respects, it is foreseeable that they could comply under certain circumstances, especially if used alongside human soldiers. Even so, there is widespread concern that allowing LARs to kill people may denigrate the value of life itself. Tireless war machines, ready for deployment at the push of a button, pose the danger of permanent (if low-level) armed conflict, obviating the opportunity for post-war reconstruction. The onus is on those who wish to deploy LARs to demonstrate that specific uses should in particular circumstances be permitted. Given the far-reaching implications for protection of life, considerable proof will be required.

---

1 Christof Heyns (2013, April 9). A/HRC/23/47. Retrieved from https://www.ohchr.org/Documents/HRBodies/HRCouncil/RegularSession/Session23/A-HRC-23-47_en.pdf

110. If left too long to its own devices, the matter will, quite literally, be taken out of human hands. Moreover, coming on the heels of the problematic use and contested justifications for drones and targeted killing, LARs may seriously undermine the ability of the international legal system to preserve a minimum world order.

111. Some actions need to be taken immediately, while others can follow afterwards. If the experience with drones is an indication, it will be important to ensure that transparency, accountability and the rule of law are placed on the agenda from the start. Moratoria are needed to prevent steps from being taken that may be difficult to reverse later, while an inclusive process to decide how to approach this issue should occur simultaneously at the domestic, intra-State, and international levels.

112. To initiate this process an international body should be established to monitor the situation and articulate the options for the longer term. The ongoing engagement of this body, or a successor, with the issues presented by LARs will be essential, in view of the constant evolution of technology and to ensure protection of the right to life – to prevent both individual cases of arbitrary deprivation of life as well as the devaluing of life on a wider scale."

Appendix C

Excerpts from the Final Document of the Fifth Review Conference – Decision to Formally Establish the CCW GGE

Excerpts from the Final Document of the Fifth Review Conference of the High Contracting Parties to the Convention on Prohibitions or Restrictions on the Use of Certain Conventional Weapons Which May Be Deemed to Be Excessively Injurious or to Have Indiscriminate Effects (2016, December 23). CCW/CONF.V/10

"9. The Conference welcomes the informal discussions held in the framework of the informal meetings of experts on emerging technologies in the area of lethal autonomous weapons systems (LAWS) in 2014, 2015 and 2016 and takes note of the reports of the Chairpersons submitted under their own responsibility

(…)

The Conference takes the following decisions:
Decision 1 To establish an open-ended Group of Governmental Experts (GGE) related to emerging technologies in the area of lethal autonomous weapons systems (LAWS) in the context of the objectives and purposes of the Convention, which shall meet for a period of ten days in 2017, adhering to the agreed recommendations contained in document CCW/CONF.V/2, and to submit a report to the 2017 Meeting of the High Contracting Parties to the Convention consistent with those recommendations.

(…)

Review
Article 8
49. Agrees to establish an open-ended Group of Governmental Experts related to emerging technologies in the area of lethal autonomous weapons systems (LAWS), which shall meet for a period of ten days in 2017, adhering to the agreed recommendations contained in document CCW/CONF.V/2, and to submit a report to the 2017 Meeting of the High Contracting Parties to the Convention consistent with those recommendations."

Appendix D

Excerpts from the Report of the 2017 Group of Governmental Experts on Lethal Autonomous Weapons Systems (LAWS)

Excerpts of the Report of the 2017 session of the Group of Governmental Experts on Emerging Technologies in the Area of Lethal Autonomous Weapons Systems CCW/GGE.1/2017/3 Convention on Prohibitions or Restrictions on the Use of Certain Conventional Weapons Which May Be Deemed to Be Excessively Injurious or to Have Indiscriminate Effects (2017, December 22)

"III. Conclusions and recommendations of the Group of Governmental Experts

16. In pursuit of its mandate, the Group affirmed that: (a) CCW offers an appropriate framework for dealing with the issue of emerging technologies in the area of lethal autonomous weapons systems. The Convention's modular and evolutionary character, the balance it seeks to strike between humanitarian considerations and military necessity as well as the opportunity it offers to engage multiple stakeholders make it an ideal platform for reaching a common understanding on this complex subject; (b) International humanitarian law continues to apply fully to all weapons systems, including the potential development and use of lethal autonomous weapons systems; (c) Responsibility for the deployment of any weapons system in armed conflict remains with States. States must ensure accountability for lethal action by any weapon system used by the State's forces in armed conflict in accordance with applicable international law, in particular international humanitarian law. The human element in the use of lethal force should be further considered; (d) Acknowledging the dual nature of technologies in the area of intelligent autonomous systems that continue to develop rapidly, the Group's efforts in the context of its mandate should not hamper progress in or access to civilian research and development and use of these technologies; (e) Given the pace of technology development and uncertainty regarding the pathways for the emergence of increased autonomy, there would be a need to keep potential military applications of related technologies under review in the context of the Group's work; (f) Keeping in mind the discussion on the various dimensions of emerging technologies in the area of lethal autonomous weapon systems – technological, military, legal and ethical, there would be merit in focusing the next stage of the Group's discussions on the characterization of the systems under consideration in order to promote a common understanding on concepts and characteristics relevant to the objectives and purposes of the CCW; (g) There is a need to further assess the aspects of human-machine

interaction in the development, deployment and use of emerging technologies in the area of lethal autonomous weapons systems in the next stage of the Group's work; (h) Further, there is need to continue the discussion in a focused and participative manner on possible options for addressing the humanitarian and international security challenges posed by emerging technologies in the area of lethal autonomous weapons systems in the context of the objectives and purposes of the Convention without prejudging policy outcomes and taking into account past, present and future proposals."

Appendix E

Excerpts from the Report of the 2018 Group of Governmental Experts on Lethal Autonomous Weapons Systems (LAWS)

Excerpts of the Report of the 2018 session of the Group of Governmental Experts on Emerging Technologies in the Area of Lethal Autonomous Weapons Systems CCW/GGE.1/2018/3 Convention on Prohibitions or Restrictions on the Use of Certain Conventional Weapons Which May Be Deemed to Be Excessively Injurious or to Have Indiscriminate Effects (2018, October 23)

"Human element in the use of lethal force; aspects of human-machine interaction in the development, deployment and use of emerging technologies in the area of lethal autonomous weapons systems

23. In the context of the objectives and purposes of the CCW, it was noted that the nature and quality of the human-machine interface is important to address concerns related to the development, deployment and use of emerging technologies in the area of lethal autonomous weapons systems. In line with the Chair's 'sunrise slide', the following touch points in the human-machine interface were considered: (0) political direction in the pre-development phase; (1) research and development; (2) testing, evaluation and certification; (3) deployment, training, command and control; (4) use and abort; (5) post-use assessment. It was noted that: (a) Accountability threads together these various human-machine touch points in the context of the CCW. Humans must at all times remain accountable in accordance with applicable international law for decisions on the use of force. (b) Where feasible and appropriate, inter-disciplinary perspectives must be integrated in research and development, including through independent ethics reviews bearing in mind national security considerations and restrictions on commercial proprietary information. (c) Weapons systems under development, or modification which significantly changes the use of existing weapons systems, must be reviewed as applicable to ensure compliance with IHL. (d) Where feasible and appropriate, verifiability and certification procedures covering all likely or intended use scenarios must be developed, the experience of applying such procedures should be shared bearing in mind national security considerations or commercial restrictions on proprietary information. (e) Accountability for the use of force in armed conflict must be ensured in accordance with applicable international law, including through the operation of any emerging weapons systems within a responsible chain of command and

control. (f) Human responsibility for the use of force must be retained. To the extent possible or feasible, this could extend to intervention in the operation of a weapon if necessary to ensure compliance with IHL. (g) Necessary investments in human resources and training should be made in order to comply with IHL and retain human accountability and responsibility throughout the development and deployment cycle of emerging technologies. (h) Keeping in mind the foregoing, and recognizing the authority and responsibility of States in this area, it would be useful to continue discussions on reaching shared understandings on the extent and quality of the human-machine interaction in the various phases of the weapons system's life cycle as well as clarifying the accountability threads throughout these phases."

Appendix F

Excerpts from the Report of the 2019b Group of Governmental Experts on Lethal Autonomous Weapons Systems (LAWS)

Excerpts of the Report of the 2019 session of the Group of Governmental Experts on Emerging Technologies in the Area of Lethal Autonomous Weapons Systems CCW/GGE.1/2019/3Convention on Prohibitions or Restrictions on the Use of Certain Conventional Weapons Which May Be Deemed to Be Excessively Injurious or to Have Indiscriminate Effects (2019, September 25)

"Conclusions

16. The Group took into consideration the guiding principles affirmed by the Group in 2018, as contained in paragraph 21 of CCW/GGE.1/2018/3, and used the principles as a basis for their work in 2019 (…)

17. On the agenda item 5 (a) "An exploration of the potential challenges posed by emerging technologies in the area of lethal autonomous weapons systems to International Humanitarian Law" the Group concluded as follows: (…) (b) IHL imposes obligations on States, parties to armed conflict and individuals, not machines; (c) States, parties to armed conflict and individuals remain at all times responsible for adhering to their obligations under applicable international law, including IHL. States must also ensure individual responsibility for the employment of means or methods of warfare involving the potential use of weapons systems based on emerging technologies in the area of lethal autonomous weapons systems in accordance with their obligations under IHL; (d) The IHL requirements and principles including *inter alia* distinction, proportionality and precautions in attack must be applied through a chain of responsible command and control by the human operators and commanders who use weapons systems based on emerging technologies in the area of lethal autonomous weapons systems; (e) Human judgment is essential in order to ensure that the potential use of weapons systems based on emerging technologies in the area of lethal autonomous weapons systems is in compliance with international law, and in particular IHL; (f) Compliance with the IHL requirements and principles, including inter alia distinction, proportionality and precautions in attack, in the potential use of weapons systems based on emerging technologies in the area of lethal autonomous weapons systems requires inter alia that human beings make certain judgments in good faith based on their assessment of the information available to them at the time; (g) In cases involving weapons systems based

on emerging technologies in the area of lethal autonomous weapons systems not covered by the CCW and its annexed Protocols or by other international agreements, the civilian population and the combatants shall at all times remain under the protection and authority of the principles of international law derived from established custom, from the principles of humanity and from the dictates of public conscience; (h) A weapons system based on emerging technologies in the area of lethal autonomous weapons systems, must not be used if it is of a nature to cause superfluous injury or unnecessary suffering, or if it is inherently indiscriminate, or is otherwise incapable of being used in accordance with the requirements and principles of IHL; (i) Legal reviews, at the national level, in the study, development, acquisition or adoption of a new weapon, means or method of warfare are a useful tool to assess nationally whether potential weapons systems based on emerging technologies in the area of lethal autonomous weapons systems would be prohibited by any rule of international law applicable to that State in all or some circumstances. States are free to independently determine the means to conduct legal reviews although the voluntary exchange of best practices could be beneficial, bearing in mind national security considerations or commercial restrictions on proprietary information."

Appendix G

AWS Guiding Principles (*CCW/MSP/2019/9*)

Meeting of the High Contracting Parties to the Convention on Prohibitions or Restrictions on the Use of Certain Conventional Weapons Which May Be Deemed to Be Excessively Injurious or to Have Indiscriminate Effects (2019, December 13). Annex III, Retrieved from https://daccess-ods.un.org/access.nsf/Get?OpenAgent &DS=CCW/MSP/2019/9&Lang=E

Guiding Principles affirmed by the Group of Governmental Experts on Emerging Technologies in the Area of Lethal Autonomous Weapons System

"It was affirmed that international law, in particular the United Nations Charter and International Humanitarian Law (IHL) as well as relevant ethical perspectives, should guide the continued work of the Group. Noting the potential challenges posed by emerging technologies in the area of lethal autonomous weapons systems to IHL, the following were affirmed, without prejudice to the result of future discussions:

(a) International humanitarian law continues to apply fully to all weapons systems, including the potential development and use of lethal autonomous weapons systems;
(b) Human responsibility for decisions on the use of weapons systems must be retained since accountability cannot be transferred to machines. This should be considered across the entire lifecycle of the weapons system;
(c) Human-machine interaction, which may take various forms and be implemented at various stages of the life cycle of a weapon, should ensure that the potential use of weapons systems based on emerging technologies in the area of lethal autonomous weapons systems is in compliance with applicable international law, in particular IHL. In determining the quality and extent of human-machine interaction, a range of factors should be considered including the operational context, and the characteristics and capabilities of the weapons system as a whole;
(d) Accountability for developing, deploying and using any emerging weapons system in the framework of the CCW must be ensured in accordance with applicable international law, including through the operation of such systems within a responsible chain of human command and control;
(e) In accordance with States' obligations under international law, in the study, development, acquisition, or adoption of a new weapon, means or method of warfare, determination must be made whether its employment would, in some or all circumstances, be prohibited by international law;

(f) When developing or acquiring new weapons systems based on emerging technologies in the area of lethal autonomous weapons systems, physical security, appropriate non-physical safeguards (including cyber-security against hacking or data spoofing), the risk of acquisition by terrorist groups and the risk of proliferation should be considered;
(g) Risk assessments and mitigation measures should be part of the design, development, testing and deployment cycle of emerging technologies in any weapons systems;
(h) Consideration should be given to the use of emerging technologies in the area of lethal autonomous weapons systems in upholding compliance with IHL and other applicable international legal obligations;
(i) In crafting potential policy measures, emerging technologies in the area of lethal autonomous weapons systems should not be anthropomorphized;
(j) Discussions and any potential policy measures taken within the context of the CCW should not hamper progress in or access to peaceful uses of intelligent autonomous technologies;
(k) The CCW offers an appropriate framework for dealing with the issue of emerging technologies in the area of lethal autonomous weapons systems within the context of the objectives and purposes of the Convention, which seeks to strike a balance between military necessity and humanitarian considerations."

Appendix H

Excerpts of the Chairperson's Summary of the 2020 Work of the GGE – Impact of the Global Covid-19 Pandemic

Excerpts from the Chairperson's Summary of the 2020 work of the Group of Governmental Experts on Emerging Technologies in the Area of Lethal Autonomous Weapons Systems CCW/GGE.1/2020/WP.7 (2021, April 2019)

"III. Conclusions

15. The Group explored and sought agreement on possible recommendations on options related to emerging technologies in the area of lethal autonomous weapons systems, in the context of the objectives and purposes of the Convention, taking into account all proposals (past, present and future) and the agenda items as reflected in paragraph 11 and annex I of the Report of its 2019 session. In its discussions under each agenda item the Group considered the legal, technological and military aspects and the interaction between them, and bearing in mind ethical considerations.

16. Delegations presented different options related to emerging technologies in the area of LAWS in the context of the objectives and purposes of the Convention as set out in paragraph 28 of the report of the Group in 2018 (CCW/GGE.1/2018/3), including a legally binding instrument, a political declaration, clarity on the implementation of existing obligations of international law in particular international humanitarian law, and the option that no further legal measures are needed. Other non-legally binding instruments were also presented. Their pros and cons were discussed.

17. Following the paragraph 31 of the 2019 Meeting of the High Contracting Parties final document (CCW/MSP/2019/9) the Group considered: (1) the guiding principles, which the Meeting of High Contracting Parties endorsed in 2019, which may be further developed and elaborated, (2) the work on the legal, technological and military aspects and (3) the conclusions of the Group, as reflected in its reports of 2017, 2018 and 2019; as well as all proposals past and present; and it used them as a basis for its work in 2021 and its consensus recommendations in relation to the clarification, consideration and development of aspects of the normative and operational framework on emerging technologies in the area of lethal autonomous weapons systems, in accordance with Decision

1 of the Fifth Review Conference of the High Contracting Parties to the Convention (CCW/CONF.V/10), consistent with CCW/CONF.V/2.

18. The Group considered different proposals on how to reflect the deliberations including possible conclusions and recommendations of the Group, but no consensus was reached.

19. A summary of the discussions held during the meetings of the Group, prepared under the Chairperson's responsibility, is attached to this report as Annex III."

Appendix I

Excerpts of the Report of the 2021 Session of the GGE – Statement on State Responsibility

Excerpts from the Report of the 2021 session of the Group of Governmental Experts on Emerging Technologies in the Area of Lethal Autonomous Weapons Systems CCW/GGE.1/2021/3 (2022, February 22)

"Conclusions

1. The Group explored and sought agreement on possible recommendations on options related to emerging technologies in the area of lethal autonomous weapons systems, in the context of the objectives and purposes of the Convention, taking into account all proposals (past, present and future) and the agenda items as reflected in paragraph 11 and annex I of the Report of its 2019 session.[1] In its discussions under each agenda item the Group considered the legal, technological and military aspects and the interaction between them, and bearing in mind ethical considerations.

2. Delegations presented different options related to emerging technologies in the area of LAWS in the context of the objectives and purposes of the Convention as set out in paragraph 28 of the report of the Group in 2018 (CCW/GGE.1/2018/3), including a legally-binding instrument, a political declaration, clarity on the implementation of existing obligations of international law in particular international humanitarian law, and the option that no further legal measures are needed. Other non-legally binding instruments were also presented. Their pros and cons were discussed.

3. Following the paragraph 31 of the 2019 Meeting of the High Contracting Parties final document (CCW/MSP/2019/9) the Group considered: (1) the guiding principles, which the Meeting of High Contracting Parties endorsed in 2019[2], which may be further developed and elaborated, (2) the work on the legal, technological and military aspects and (3) the conclusions of the Group, as reflected in its reports of 2017, 2018 and 2019; as well as all proposals past and present; and it used them as a basis for its work in 2021 and its consensus recommendations in relation to the clarification, consideration and development of aspects of the normative and operational framework on emerging technologies in the area of lethal autonomous weapons systems, in accordance with Decision

[1] CCW/GGE.1/2019/3.

[2] CCW/GGE.1/2019/3, Annex III.

1 of the Fifth Review Conference of the High Contracting Parties to the Convention (CCW/CONF.V/10), consistent with CCW/CONF.V/2.

4. The Group considered different proposals on how to reflect the deliberations including possible conclusions and recommendations of the Group, but no consensus was reached.

5. A summary of the discussions held during the meetings of the Group, prepared under the Chairperson's responsibility, is attached to this report as Annex III."

Appendix J

Excerpts of the 2021 Sixth Review Conference of the High Contracting Parties to the Convention on Prohibitions or Restrictions on the Use of Certain Conventional Weapons Which May Be Deemed to Be Excessively Injurious or to Have Indiscriminate Effects

Excerpts from the Final document of the 6th Review Conference of the High Contracting Parties to the Convention on Prohibitions or Restrictions on the Use of Certain Conventional Weapons Which May Be Deemed to Be Excessively Injurious or to Have Indiscriminate Effects CCW/CONF.VI/11 (2022, January 10).

"IV. Work of the Sixth Review Conference

(…)

1. The Chair of the 2021 Group of Governmental Experts on Emerging Technologies in the area of Lethal Autonomous Weapons Systems presented its report to be considered by the Conference. The Chair of the 2021 Group of Governmental Experts on Emerging Technologies in the area of Lethal Autonomous Weapons Systems, under his own responsibility and initiative, has prepared a Chairperson's summary contained in annex III of the document CCW/GGE.1/2021/CRP.1. The Conference noted that this paper had not been agreed and had no status. It was the Chair's view that the Chairperson's summary could assist delegations and could constitute a useful basis for future work of the Group of Governmental Experts on Emerging Technologies in the area of Lethal Autonomous Weapons Systems.

"The Conference takes the following decisions:

Decision 1

The Sixth Review Conference decides that the work of the open-ended Group of Governmental Experts related to emerging technologies in the area of lethal autonomous weapon systems established by Decision 1 of the Fifth Review Conference as contained in document CCW/CONF.V/10, adhering to the agreed recommendations contained in document CCW/CONF.V/2, is to continue, to strengthen the Convention.

In the context of the objectives and purpose of the Convention, the Group is to consider proposals and elaborate, by consensus, possible measures, including taking into account the example of existing protocols within the Convention, and other options related to the normative and operational framework on emerging technologies in the area of lethal autonomous weapon systems, building upon the recommendations and conclusions of the Group of Governmental Experts related to emerging technologies in the area of lethal autonomous weapon systems, and bringing in expertise on legal, military, and technological aspects.

International law, in particular the United Nations Charter and international humanitarian law, as well as relevant ethical perspectives should guide the continued work of the Group.

The rules of procedure of the Review Conference shall apply mutatis mutandis to the Group.

The Group shall conduct its work and adopt its report by consensus and shall submit a report to the meeting of High Contracting Parties. The widest possible participation of all High Contracting Parties is to be promoted in accordance with the goals of the CCW Sponsorship Program.

The Group shall meet for 10 days in Geneva in 2022.

The GGE Chair will be designated by consensus through a written silence procedure."

Appendix K

Excerpts of the Report of the 2022 Session of the GGE – Statement on State Responsibility and Comparative Table (Chair's Draft Final Report x Report)

K.1 Excerpts from the Report of the 2022 session of the Group of Governmental Experts on Emerging Technologies in the Area of Lethal Autonomous Weapons Systems CCW/GGE.1/2022/2 (2022, August 31)

Convention on Prohibitions or Restrictions on the Use of Certain Conventional Weapons Which May Be Deemed to Be Excessively Injurious or to Have Indiscriminate Effects (2022, August 31)

"III. Conclusions

16. The Group recalled Decision 1 by the Sixth Review Conference contained in document CCW/CONF.VI/11, that the work of the open-ended Group of Governmental Experts related to emerging technologies in the area of lethal autonomous weapon systems established by Decision 1 of the Fifth Review Conference as contained in document CCW/CONF.V/10, adhering to the agreed recommendations contained in document CCW/CONF.V/2, is to continue, to strengthen the Convention. In accordance with its mandate, in the context of the objectives and purpose of the Convention, the Group considered proposals and discussed the elaboration, by consensus, of possible measures, including taking into account the example of existing protocols within the Convention, and other options related to the normative and operational framework on emerging technologies in the area of LAWS, building upon the recommendations and conclusions of the Group of Governmental Experts related to emerging technologies in the area of lethal autonomous weapon systems, and bringing in expertise on legal, military, and technological aspects.

17. In this regard, delegations presented and discussed a number of proposals on possible measures and options including: legally binding instruments under the framework of the CCW; a non-legally binding instrument; clarity on the implementation of existing obligations under international law, in particular IHL; an option that prohibits and regulates on the basis of IHL; and the option that no further legal measures are needed.

18. The Group recalled that the right of parties to an armed conflict to choose methods or means of warfare is not unlimited. Furthermore, international law, in

particular the United Nations Charter and International Humanitarian Law (IHL) as well as relevant ethical perspectives, should continue to guide the work of the Group. The Group reaffirmed the relevant paragraphs related to emerging technologies in the area of LAWS of documents CCW/CONF.VI/11 and CCW/CONF.V/10.

19. **For the purposes of its work, the Group recognized that every internationally wrongful act of a State, including those potentially involving weapons systems based on emerging technologies in the area of LAWS entails international responsibility of that State, in accordance with international law. In addition, States must comply with international humanitarian law. Humans responsible for the planning and conducting of attacks must comply with international humanitarian law. (emphasis added)**

IV. Recommendations

20. The Group recommends that:
(a) The work of the open-ended Group of Governmental Experts related to emerging technologies in the area of lethal autonomous weapon systems (…), is to continue, to strengthen the Convention. In the context of the objectives and purpose of the Convention, the Group is to intensify the consideration of proposals and elaborate, by consensus, possible measures, including taking into account the example of existing protocols within the Convention, and other options related to the normative and operational framework on emerging technologies in the area of lethal autonomous weapon systems, building upon the recommendations and conclusions of the Group of Governmental Experts related to emerging technologies in the area of lethal autonomous weapon systems, and bringing in expertise on legal, military, and technological aspects (…)"

K.2 Comparative Table of Excerpts of the Chair's Draft final report CCW/GGE.1/2022/CRP.1 and the Report of the 2022 session of the Group of Governmental Experts on Emerging Technologies in the Area of Lethal Autonomous Weapons Systems CCW/GGE.1/2022/2(2022, August 31)

Comparative Table

Considering the relevance of the 2022 report to State responsibility, and also that the author believes that the Group of Governmental Experts on Emerging Technologies in the Area of Lethal Autonomous Weapons Systems missed the opportunity of agreeing on a more substantial report, as proposed by the Chair, both regarding State responsibility and other issues, the following table is provided.

Chair's Draft Final Report CCW/GGE.1/2022/CRP.1 (2022, July 29)	**Report of the 2022 session of the Group of Governmental Experts on Emerging Technologies in the Area of Lethal Autonomous Weapons Systems CCW/ GGE.1/2022/2 (2022, August 31)**
"1. **The CCW remains the appropriate forum for dealing with the issue of emerging technologies in the area of LAWS, within the context of its objectives and purposes**. The right of parties to an armed conflict to choose methods or means of warfare, including weapons systems based on emerging technologies in the area of LAWS, is not unlimited. Furthermore, international law, in particular the United Nations Charter and International Humanitarian Law (IHL) as well as relevant ethical perspectives, should continue to guide the work of the Group."	16. **The Group recalled Decision 1 by the Sixth Review Conference contained in document CCW/CONF.VI/11, that the work of the open-ended Group of Governmental Experts related to emerging technologies in the area of lethal autonomous weapon systems established by Decision 1 of the Fifth Review Conference as contained in document CCW/CONF.V/10, adhering to the agreed recommendations contained in document CCW/CONF.V/2, is to continue, to strengthen the Convention. In accordance with its mandate, in the context of the objectives and purpose of the Convention, the Group considered proposals and discussed the elaboration, by consensus, of possible measures, including taking into account the example of existing protocols within the Convention, and other options related to the normative and operational framework on emerging technologies in the area of LAWS, building upon the recommendations and conclusions of the Group of Governmental Experts related to emerging technologies in the area of lethal autonomous weapon systems, and bringing in expertise on legal, military, and technological aspects.**
	17. **In this regard, delegations presented and discussed a number of proposals on possible measures and options including: legally binding instruments under the framework of the CCW; a non-legally binding instrument; clarity on the implementation of existing obligations under international law, in particular IHL; an option that prohibits and regulates on the basis of IHL; and the option that no further legal measures are needed.**

	18. The Group recalled that the right of parties to an armed conflict to choose methods or means of warfare is not unlimited. Furthermore, international law, in particular the United Nations Charter and International Humanitarian Law (IHL) as well as relevant ethical perspectives, should continue to guide the work of the Group. **The Group reaffirmed the relevant paragraphs related to emerging technologies in the area of LAWS of documents CCW/CONF. VI/11 and CCW/CONF.V/10.**
1. On the basis of its consideration of proposals, the Group endorsed the following elements or possible measures, including taking into account the example of existing protocols within the Convention, and other options related to the normative and operational framework on emerging technologies in the area of lethal autonomous weapon systems. 2. A weapons system based on emerging technologies in the area of LAWS must not be possessed, developed or deployed with a view towards use, or used if it: (a) is designed or used in such a way that its effects in attacks cannot be anticipated or controlled, as required in the circumstances by international humanitarian law, particularly the principles of distinction, proportionality, and precaution in attack; (b) is designed or used to conduct attacks that would not be the responsibility of the human command under which the system would be used.	
3. To address the risks of unintended engagements, incidental loss of life, injuries to civilians and damage to civilian objects resulting from the use of weapons systems based on emerging technologies in the area of LAWS, High Contracting Parties should elaborate and adopt measures across the life-cycle of the weapons system, including to:	

(a) control, limit, or otherwise affect the types of targets that the system can engage; (b) control, limit, or otherwise affect the duration, geographical scope, and scale of operation of the system, such as the incorporation of self-destruct, self-deactivation, or self-neutralization mechanisms into munitions and weapons systems;	
(c) ensure human operators are informed and empowered to make decisions and exert control over the use of force, in the circumstances of their use, to comply with international humanitarian law; (d) integrate risk assessments into the design, development, testing and deployment stages of the entire life-cycle of a weapon system; (e) reduce automation bias in systems operators as well as unintended bias in artificial intelligence capabilities related to the use of the weapon system. 4. **Compliance with international humanitarian law in the use of weapon systems based on emerging technologies in the area of LAWS requires, *inter alia,* good-faith human judgement based on the assessment of the information available at the time, and the ability of the operators to make decisions and exert control over the use of force. Moreover, humans must at all times remain accountable in accordance with applicable international law for decisions over the use of force.**	19. (…)**In addition, States must comply with international humanitarian law. Humans responsible for the planning and conducting of attacks must comply with international humanitarian law.**
5. Every internationally wrongful act of a State, including such conduct involving weapon systems based on emerging technologies in the area of LAWS entails international responsibility of that State. **The conduct of a State's organs such as its agents and all persons forming part of its armed forces, is attributable to that State**, including any such acts	19. For the purposes of its work, the Group recognized that every internationally wrongful act of a State, including those potentially involving weapons systems based on emerging technologies in the area of LAWS entails international responsibility of that State, in accordance with international law (…)

and omissions involving the use of a weapons system based on emerging technologies in the area of LAWS, in accordance with applicable international law.	
6. In accordance with States' obligations under international law, in the study, development, acquisition, or adoption of a new weapon, means or method of warfare, including such potential weapons systems based on emerging technologies in the area of LAWS, States must determine whether its employment would, in some or all circumstances, be prohibited by international law. 7. Measures related to weapons systems based on emerging technologies in the area of LAWS should address the following topics: (a) Weapons prohibited from use in all circumstances; (b) Other prohibitions or restrictions on the use; (c) Responsibility and accountability; (d) Good practices related to human-machine interaction and legal reviews; (e) Risk assessments and mitigation measures.	

Appendix L

Excerpts of the Report of the 2023 Session of the GGE – Statement on State Responsibility

Excerpts of the Report of the 2023 session of the Group of Governmental Experts on Emerging Technologies in the Area of Lethal Autonomous Weapons Systems CCW/GGE.1/2023/2

Convention on Prohibitions or Restrictions on the Use of Certain Conventional Weapons Which May Be Deemed to Be Excessively Injurious or to Have Indiscriminate Effects (2023, May 24)

"III. Conclusions

17. The Group acknowledged the value of the conclusions and recommendations of the Group and the agreed recommendations contained in document CCW/CONF.V/2, and the endorsement by the Meeting of the High Contracting Parties in 2019 of the Guiding Principles affirmed by the Group.

18. The Group recalled that the right of parties to an armed conflict to choose methods or means of warfare is not unlimited. Furthermore, international law, in particular the United Nations Charter and International Humanitarian Law (IHL) as well as relevant ethical perspectives, should continue to guide the work of the Group. The Group reaffirmed the relevant paragraphs related to emerging technologies in the area of LAWS of documents CCW/CONF.VI/11 and CCW/CONF.V/10.

19. Building upon its previous work and in accordance with its mandate, in the context of the objectives and purpose of the Convention, and without prejudice to the future work of the Group, on the basis of the intensified consideration of proposals, the Group concluded as follows:

20. The characterization of weapon systems based on emerging technologies in the area of lethal autonomous weapons systems should take into consideration the possible future development of those technologies.

21. Without prejudice to the future work of the Group that continues to be guided by international law, in particular the United Nations Charter and IHL as well as relevant ethical perspectives, the Group concluded that: (a) IHL continues to apply fully to the potential development and use of LAWS; (b) Weapons systems

based on emerging technologies in the area of LAWS must not be used if they are incapable of being used in compliance with IHL; (c) Control with regard to weapon systems based on emerging technologies in the area of LAWS is needed to uphold compliance with international law, in particular IHL, including the principles and requirements of distinction, proportionality and precautions in attack.

22. States must ensure compliance with their obligations under international law, in particular IHL, throughout the lifecycle of weapon systems based on emerging technologies in the area of LAWS. When necessary, States should, *inter alia*: (a) Limit the types of targets that the system can engage; (b) Limit the duration, geographical scope, and scale of the operation of the weapon system; (c) Provide appropriate training and instructions for human operators.

23. In accordance with States' obligations under international law, in the study, development, acquisition, or adoption of a new weapon, means or method of warfare, determination must be made whether its employment would, in some or all circumstances, be prohibited by international law. In this context, the voluntary exchange of relevant best practices between States is encouraged, bearing in mind national security considerations or commercial restrictions on proprietary information."

Appendix M

Use of AWS in Libya

Excerpts from United Nations Security Council (UNSC). (2021, March 8). Letter dated 8 March 2021 from the Panel of Experts on Libya established pursuant to K (2011) addressed to the President of the Security Council and its Annex 30. S/2021/229. Retrieved from https://undocs.org/S/2021/229

"63. On 27 March 2020, the Prime Minister, FaiezSerraj, announced the commencement of Operation PEACE STORM, 46 which moved GNA-AF to the offensive along the coastal littoral. The combination of the Gabya-class frigates and Korkut short-range air defence systems provided a capability to place a mobile air defence bubble around GNA-AF ground units, which took HAF air assets out of the military equation. The enhanced operational intelligence capability included Turkish operated signal intelligence and the intelligence, surveillance and reconnaissance provided by Bayraktar TB-2 and probably TAI Anka S unmanned combat aerial vehicles (see annex 27). This allowed for the development of an asymmetrical war of attrition designed to degrade HAF ground unit capability. The GNA-AF breakout of Tripoli was supported with Firtina T155 155mm self-propelled guns (see annex 28) and T-122 Sakarya multi-launch rocket systems (see annex 29) firing extended range precision munitions against the mid-twentieth century main battle tanks and heavy artillery used by HAF. Logistics convoys and retreating HAF were subsequently hunted down and remotely engaged by the unmanned combat aerial vehicles or the lethal autonomous weapons systems such as the STM Kargu-2 (see annex 30) and other loitering munitions. The lethal autonomous weapons systems were programmed to attack targets without requiring data connectivity between the operator and the munition: in effect, a true "fire, forget and find" capability. The unmanned combat aerial vehicles and the small drone intelligence, surveillance and reconnaissance capability of HAF were neutralized by electronic jamming from the Koral electronic warfare system.47

64. The concentrated firepower and situational awareness that those new battlefield technologies provided was a significant force multiplier for the ground units of GNA-AF, which slowly degraded the HAF operational capability. The latter's units were neither trained nor motivated to defend against the effective use of this new technology and usually retreated in disarray. Once in retreat, they were subject to continual harassment from the unmanned combat aerial vehicles and lethal autonomous weapons systems, which were proving to be a

highly effective combination in defeating the United Arab Emirates-delivered Pantsir S-1 surface-toair missile systems. These suffered significant casualties, even when used in a passive electro-optical role to avoid GNA-AF jamming. With the Pantsir S-1 threat negated, HAF units had no real protection from remote air attacks."

Appendix N

Comments on the 2023 U.S. 2023 DoD Directive 30009 on Autonomy in Weapon Systems

Article published by the author commenting the 2023 U.S. 2023 DoD Directive 30009 on Autonomy in Weapon Systems – Published at Brazilian Journal of International Affairs Year 2 / N° 7 / Jul-Set 2023. Available at https://cebri-revista.emnuvens.com.br/revista/article/view/153/210

Exploring the 2023 U.S. Directive on Autonomy in Weapon Systems: Key Advancements and Potential Implications for International Discussions

Lutiana Valadares Fernandes Barbosa

In 2012 the United States (U.S.) published the Department of Defense (DoD) Directive 3000.09, "Autonomy in Weapon Systems" (United States Department of Defense 2012). By then, the debate over Autonomous Weapons Systems (AWS) was at a very initial state. Three years later, the topic began to be informally discussed at the United Nations (UN) and, in 2017, formally discussed by The Group of Governmental Experts (GGE) on AWS under the auspices of the Convention on Certain Conventional Weapons (CCW) (UNODA 2023).

Throughout the last decade, AWS, which initially resembled characters of science fiction movies, have been used in the international scenario.[1] In 2021, an expert panel report addressed to the UN Security Council acknowledged that the attack drone AWS STM Kargu-2 (Kargu 2022) was deployed in Libya in 2020. According to the report "The lethal autonomous weapons systems were programmed to attack targets without requiring data connectivity between the operator and the munition: in effect, a true 'fire, forget and find' capability" (UNSC 2021, 17). It is also argued that AWS may have been used in the ongoing Russian–Ukraine conflict (Kallenborn 2022).

---

[1] Note that since there is no internationally agreed definition of AWS, there are distinctive views regarding their use or not. See for example:
Scharre (2018, 46): "Generally speaking, fully autonomous weapons are not in wide use, but there are a few select systems that cross the line" and
Crootof (2015, 1837): "With this new definition it quickly becomes clear that, contrary to the general consensus, autonomous weapon systems are not weapons of the future: they exist and are in use today." See also (UK 2017–2018, 14): "The UK does not possess fully autonomous weapon systems and has no intention of developing them. Such systems are not yet in existence and are not likely to be for many years, if at all."

Despite the excellent work of the Brazilian Chair of the 2021–2023 CCW GGE and predecessors, diplomatic efforts and achievements, diplomatic pace is much slower than technological development. While AWS remain without specific regulation in the international arena, few States – such as the U.S. and the United Kingdom (United Kingdom Ministry of Defense 2017) – have their directives on AWS, or have made them publicly available. The 2012 U.S. Directive 3000.09 has not only expressed the U.S. position on the debate, but also impacted the international discussions, as it was the first State directive on AWS (Insinna and Mehta 2022) and reflected in the U.S. delegation statements at the CCW GGE (United States Delegation 2018, p. 2). Its concept, for instance, was embraced by international NGOs (Horowitz 2016, p. 85).

The 2012 Directive 3000.09 was published before the debate at the UN began, but has highly influenced international discussions. Following the technological developments since then, the DoD published in 2023 a new Directive 3000.09 with the potential to impact the ongoing international debate, which takes place mainly under the auspices of the CCW. The novel definition, for instance, is the one contained in Australia, Canada, Japan, the Republic of Korea, the United Kingdom, and the United States delegations, Draft articles on AWS prohibitions and other regulatory measures on the basis of International Humanitarian Law (IHL), submitted to the CCW GGE in 2023 (Australia et al. 2023).

The present article aims to answer the questions of which were the main novelties of the 2023 Directive 3000.09 and which are its possible impacts on the international discussion on AWS. To answer this question, qualitative research was developed. Bibliographical research was conducted regarding AWS. Its results were analyzed using the hypothetical-deductive method. A documentary survey was carried out in the DoD Directives and UN documents. Its results were analyzed using the inductive method.

The article first critically discusses the 2023 DoD Directive 3000.09. Its main novelties are, substituting the word "human operator" for "operator" in the definition of AWS and semi-AWS, adding a restriction on the requirement of appropriate levels of human judgment and elucidating the definition of failure, adding new requirements for AWS review and deployment, introducing concepts such as transparency, auditability, and explainability, and establishing the AWS Working Group. After critically presenting the main shifts, in the final considerations, this article discusses the good practices and pushbacks to the international community.

Less than one month after the 2023 Directive was issued, the U.S. launched a Political Declaration at the Responsible AI in the Military Domain (REAIM) Conference in the Hague (United States Department of State, 2023). The political declaration enunciates what the U.S. envisions as best practices and shared values on responsible use of AI in the military domain and calls on States to adhere to it. Therefore, while considering the pros and cons of the innovations of the 2023 DoD Directive 3000.09 we will do so considering the U.S. Political Declaration on Responsible Military Use of Artificial Intelligence and Autonomy.

Bibliography

Abbott, R. (2020). The Reasonable Robot: Artificial Intelligence and the Law. Cambridge University Press.

Acheson, R. (2021, August 8). The Human Element. Civil society perspectives on the Convention on Certain Conventional Weapons (CCW)'s Group Governmental Experts on Lethal Autonomous Weapon Systems 3–13 August 2021. CCW Report, 9(3). Reaching Critical Will.

Adankon, M. and Cheriet, M. (2009). Support Vector Machine. *In:* Li, S.Z., Jain, A. (Eds.), Encyclopedia of Biometrics. Springer, Boston, MA. https://doi.org/10.1007/978-0-387-73003-5_299 Accessed 13 November 2022.

Alexander, L. and Moore, M. (2021). Deontological Ethics. *In:* E.N. Zalta (Ed.), The Stanford Encyclopedia of Philosophy. https://plato.stanford.edu/archives/win2021/entries/ethics-deontological/ Accessed 16 November 2022.

Ali, S. (2020). Coming to a Battlefield Near You: Quantum Computing, Artificial Intelligence, & Machine Learning's Impact on Proportionality. Santa Clara J. Int'l L., 18, 1.

Allen, G.C. (2019, February 6). Understanding China's AI Strategy Clues to Chinese Strategic Thinking on Artificial Intelligence and National Security. CNAS. https://www.cnas.org/publications/reports/understanding-chinas-ai-strategy accessed 1 October 2022.

Allen, G.C. (2022, May 26). [Tweet] accessed 8 July 2022 https://twitter.com/Gregory_C_Allen/status/1529819655736643584 Accessed 12 November 2022.

Amoroso, D. (2017). Jus in bello and jus ad bellum arguments against autonomy in weapons systems: A re-appraisal. QIL, Zoom-in, 43, 21-22. http://www.qil-qdi.org/wp-content/uploads/2017/10/02_AWS_Amoroso_FIN-2.pdf Accessed 11 November 2018.

Amoroso, D. (2020). Autonomous Weapons Systems and International Law: A Study on Human-Machine Interactions in Ethically and Legally Sensitive Domains. Edizioni Scientifiche Italiane.

Anand, A. and Puscas, I. (2022, July 18). Proposals Related to Emerging Technologies in the Area of Lethal Autonomous Weapons Systems: A Resource Paper. UNIDIR (United Nations Institute for Disarmament Research). https://www.unidir.org/sites/default/files/2022-08/UNIDIRProposals_Emerging_Technologies_Lethal_Autonomous_Weapons_Systems.pdf. Accessed November 14, 2022.

Anderson, K. and Waxman, M.C. (2013). Law and Ethics for Autonomous Weapon Systems: Why a Ban Won't Work and How the Laws of War Can. Stanford University, The Hoover Institution Jean Perkins Task Force on National Security & Law Essay Series. https://scholarship.law.columbia.edu/cgi/viewcontent.cgi?article=2804&context=faculty_scholarship Accessed 16 November 2022.

Antebi, L. (2018). The International Process to Limit Autonomous Weapon Systems: Significance for Israel. Strategic Assessment, 21. https://www.inss.org.il/wp-content/uploads/2018/11/Antebi.pdf Accessed 24 August 2023.

Argentina, Costa Rica, Guatemala, Kazakhstan, Nigeria, Panama, the Philippines, Sierra Leone, State of Palestine, and Uruguay's delegation. (2022a). Proposal: Roadmap Towards New Protocol on Autonomous Weapons Systems. https://meetings.unoda.org/meeting/57989/documents?f%5B0%5D=author_documents_%3AMultiple%20States. Accessed November 11, 2022.

Argentina, Costa Rica, Ecuador, Nigeria, Panama, the Philippines, Sierra Leone, and Uruguay's delegation. (2022b). Draft Protocol VI. https://documents.unoda.org/wp-content/uploads/2022/07/WP-Argentina_CostaRica_Ecuador_Nigeria_Panama_Philippines_Sierra-Leone_Uruguay.pdf. Accessed November 11, 2022.

Argentina, Costa Rica, Ecuador, Guatemala, Kazakhstan, Nigeria, Panama, Peru, the Philippines, Sierra Leone, State of Palestine, Uruguay delegations. (2022c). Written Commentary by the Delegations of Argentina, Costa Rica, Ecuador, Guatemala, Kazakhstan, Nigeria, Panama, Peru, the Philippines, Sierra Leone, State of Palestine, Uruguay, Calling for a Legally-Binding Instrument on Autonomous Weapon Systems. https://meetings.unoda.org/meeting/57989/documents?f%5B0%5D=author_documents_%3AMultiple%20States. Accessed November 11, 2022.

Argentina, Austria, Belgium, Chile, Costa Rica, Ecuador, Guatemala, Ireland, Kazakhstan, Liechtenstein, Luxembourg, Malta, Mexico, New Zealand, Nigeria, Panama, Peru, the Philippines, Sierra Leone, Sri Lanka, State of Palestine, Switzerland, and Uruguay delegations. (2022d). Working Paper submitted to the 2022 Chair of the Group of Governmental Experts (GGE) on emerging technologies in the area of lethal autonomous weapons systems (LAWS) on behalf of Argentina, Austria, Belgium, Chile, Costa Rica, Ecuador, Guatemala, Ireland, Kazakhstan, Liechtenstein, Luxembourg, Malta, Mexico, New Zealand, Nigeria, Panama, Peru, the Philippines, Sierra Leone, Sri Lanka, State of Palestine, Switzerland, and Uruguay. https://documents.unoda.org/wp-content/uploads/2022/05/2022-GGE-LAWS-joint-submission-working-paper-G-23.pdf. Accessed November 16, 2022.

Arkin, R. (2009). Governing Lethal Behavior in Autonomous Robots. CRC Press.

Asaro, P. (2014). Ethical Issues Raised by Autonomous Weapon Systems. *In:* Autonomous Weapon Systems: Technical, Military, Legal and Humanitarian Aspects. Expert meeting (ICRC, 26-28 March 2014). ReliefWeb. https://reliefweb.int/sites/reliefweb.int/files/resources/4221-002-autonomous-weapons-systems-full-report%20%281%29.pdf Accessed 14 November 2022.

Asaro, P. (2016). Jus Nascendi: Robotic Weapons and the Martens Clause. *In:* R. Calo, M. Froomkin, & I. Kerr (Eds.), Robot Law. Edward Elgar Publishing.

Atherton, K. (2021, August 4). Loitering munitions preview the autonomous future of warfare. Brookings Tech Stream. https://www.brookings.edu/techstream/loitering-munitions-preview-the-autonomous-future-of-warfare/ Accessed September 19, 2023.

Australia, Canada, Japan, the Republic of Korea, the United Kingdom, and the United States delegations. (2022). Principles and Good Practices on Emerging Technologies in the Area of Lethal Autonomous Weapons Systems - Proposed by Australia, Canada, Japan, the Republic of Korea, the United Kingdom, and the United States. March 7, 2022. https://meetings.unoda.org/meeting/57989/documents?f%5B0%5D=author_documents_%3AMultiple%20States. Accessed November 11, 2022.

Austria, Belgium, Brazil, Chile, Ireland, Germany, Luxembourg, Mexico, and New Zealand delegations. (2020). Joint 'Commentary' on Guiding Principles A, B, C and D submitted by Austria, Belgium, Brazil, Chile, Ireland, Germany, Luxembourg, Mexico, and New Zealand. https://documents.unoda.org/wp-content/uploads/2020/09/GGE20200901-Austria-Belgium-Brazil-Chile-Ireland-Germany-Luxembourg-Mexico-and-New-Zealand.pdf. Accessed November 15, 2022.

Barbier, S. (2010). Assurances and guarantees of non-repetition - Part IV: The Content of International Responsibility. *In:* J. Crawford, et al. (Eds.), The Law of International Responsibility. Oxford University Press.

Barbosa, L. (2023). Exploring the 2023 U.S. Directive on Autonomy in Weapon Systems: Key Advancements and Potential Implications for International Discussions. CEBRI-Journal, Year 2, No. 7, 117-136. https://cebri-revista.emnuvens.com.br/revista/article/view/153/210

Barbosa, L. and Macedo, G. (2022, August 29). Should we be optimistic about the recent UN talks on killer robots? Pass Blue. https://www.passblue.com/2022/08/29/should-we-be-optimistic-about-the-recent-un-talks-on-killer-robots/ Accessed 5 September 2022.

Barnidge, R. (2006). The due diligence principle under international law. International Community Law Review, 8, 81.

Bathaee, Y. (2018). The artificial intelligence black box and the failure of intent and causation. Harvard Journal of Law & Technology, 31, 889.

Berkes, A. (2018). The standard of 'Due Diligence' as a result of interchange between the law of armed conflict and general international law. Journal of Conflict and Security Law, 23(3), 433-460.

Bhuta, N., Beck, S., Geiβ, R., Liu, H. and Kreβ, C. (2016). Autonomous Weapons Systems: Law, Ethics, Policy. Cambridge University Press.

Bikeev, I. and others. (2019). Criminological Risks and Legal Aspects of Artificial Intelligence Implementation. *In:* Proceedings of the International Conference on Artificial Intelligence, Information Processing and Cloud Computing - AIIPCC '19 (ACM Press 2019). Retrieved from http://dl.acm.org/citation.cfm?doid=3371425.3371476 Accessed 17 November 2022.

Bjorge, E. (2014, December 15). Introducing the evolutionary interpretation of treaties. EJIL Talk. https://www.ejiltalk.org/introducing-the-evolutionary-interpretation-of-treaties/ Accessed 17 November 2022.

Bo, M., Bruun, L. and Boulanin, V. (2022). Retaining Human Responsibility in the Development and Use of Autonomous Weapons Systems: On Accountability for Violations of International Humanitarian Law Involving AWS. SIPRI.

Bode, I. and Huelss, H. (2018). Autonomous weapons systems and changing norms in international relations. Review of International Studies, 44(3), 393-413. doi:10.1017/S0260210517000614.

Bolivarian Republic of Venezuela on behalf of the Non-Aligned Movement (NAM) and Other States Parties to the Convention on Certain Conventional Weapons (CCW). (2020, September 14). Working Paper by the Bolivarian Republic of Venezuela on behalf of the Non-Aligned Movement (NAM) and Other States Parties to the Convention on Certain Conventional Weapons (CCW) (CCW/GGE.1/2020/WP.5). United Nations Documents System. https://documents-dds-ny.un.org/doc/UNDOC/GEN/G20/229/09/PDF/G2022909.pdf?OpenElement. Accessed November 14, 2022.

Bolivarian Republic of Venezuela on behalf of the Non-Aligned Movement (NAM) and Other States Parties to the Convention on Certain Conventional Weapons (CCW). (2022, August 1). Working Paper by the Bolivarian Republic of Venezuela on behalf of the Non-Aligned Movement (NAM) and Other States Parties to the Convention on Certain Conventional Weapons (CCW) 7-11 March and 25-29 July 2022. https://documents.unoda.org/wp-content/uploads/2022/08/WP-NAM.pdf. Accessed November 15, 2022.

Bordin, F.L. (2018). The Analogy between States and International Organizations (Cambridge Studies in International and Comparative Law). Cambridge: Cambridge University Press. doi:10.1017/9781316658963.

Bordin, F.L. (2021, August 3). Still going strong: Twenty years of the articles on state responsibility's 'Paradoxical' relationship between form and authority. EJIL Talk! https://www.ejiltalk.org/still-going-strong-twenty-years-of-the-articles-on-state-responsibilitys-paradoxical-relationship-between-form-and-authority/ Accessed 15 November 2022.

Boulanin, V. and Verbruggen, M. (2017). Mapping the development of autonomy in weapon systems. SIPRI. https://www.sipri.org/sites/default/files/2017-11/siprireport_mapping_the_development_of_autonomy_in_weapon_systems_1117_1.pdf Accessed 19 October 2018.

Brazilian Delegation. (2020a). LAWS and Human Control: Brazilian Proposals for Working Definitions (CCW/GGE.1/2020/WP.4). Group of Governmental Experts on Emerging Technologies in the Area of Lethal Autonomous Weapons Systems, Geneva, September 21-25 and November 2-6, 2020. United Nations Documents System. https://documents-dds-ny.un.org/doc/UNDOC/GEN/G20/212/17/PDF/G2021217.pdf?OpenElement. Accessed November 11, 2022.

Brazilian Delegation. (2020b). Operationalizing the Guiding Principles: A Roadmap for the GGE on LAWS (CCW/GGE.1/2020/WP.3). Group of Governmental Experts on Emerging Technologies in the Area of Lethal Autonomous Weapons Systems, Geneva, September 21-25 and November 2-6, 2020. https://documents.unoda.org/wp-content/uploads/2020/08/CCW-GGE.1-2020-WP.3-.pdf. Accessed November 11, 2022.

Brierly, J.L. and Clapham, A. 2012. Brierly's Law of Nations: An Introduction to the Role of International Law in International Relations. Seventh edition. Oxford University Press, 2012.

Business and Corporate Litigation Committee, Business Law Section, American Bar Association. (2021, June). Recent Developments in Artificial Intelligence Cases. Business Law Today. https://businesslawtoday.org/2021/06/recent-developments-in-artificial-intelligence-cases/. Accessed June 16, 2021.

Cambridge Dictionary. (n.d.). System. Cambridge Dictionary. https://dictionary.cambridge.org/us/dictionary/english/system. Accessed November 13, 2022.

Chair of the 2020 Group of Governmental Experts (GGE) on Emerging Technologies in the Area of Lethal Autonomous Weapons Systems (LAWS). (2020). Group of Governmental Experts on Emerging Technologies in the Area of Lethal Autonomous Weapons Systems Commonalities in National Commentaries on Guiding Principles. https://documents.unoda.org/wp-content/uploads/2020/09/Commonalities-paper-on-operationalization-of-11-Guiding-Principles.pdf. Accessed November 13, 2022.

Chair of the 2021 Group of Governmental Experts (GGE) on Emerging Technologies in the Area of Lethal Autonomous Weapons Systems (LAWS). (2021, September 20). Draft Elements on Possible Consensus Recommendations in Relation to the Clarification, Consideration, and Development of Aspects of the Normative and Operational Framework on Emerging Technologies in the Area of Lethal Autonomous Weapons Systems. Revised Chair's Paper. https://reachingcriticalwill.org/images/documents/

Disarmament-fora/ccw/2021/gge/documents/chair-paper-september.pdf. Accessed November 17, 2021.

Chair of the 2022 Group of Governmental Experts (GGE) on Emerging Technologies in the Area of Lethal Autonomous Weapons Systems (LAWS). (2022). Draft Final Report (CCW/GGE.1/2022/CRP.1). https://documents.unoda.org/wp-content/uploads/2022/07/CCW-GGE.1-2022-CRP.1.docx. Accessed November 15, 2022.

Chertof, P. (2018, October). Strategic Security Analysis Perils of Lethal Autonomous Weapons Systems Proliferation: Preventing Non-State Acquisition. Geneva Centre for Security Policy. https://www.gcsp.ch/publications/perils-lethal-autonomous-weapons-systems-proliferation-preventing-non-state. Accessed November 17, 2022.

Chile and Mexico Delegations. (2022, August 1). Elements for a Legally Binding Instrument to Address the Challenges Posed by Autonomy in Weapon Systems. Working Paper submitted by Chile and Mexico. https://documents.unoda.org/wp-content/uploads/2022/08/WP-Chile-and-Mexico-.pdf. Accessed November 15, 2022.

Commission to the European Parliament, the Council, and the European Economic and Social Committee. (2020). Report on the safety and liability implications of Artificial Intelligence, the Internet of Things and robotics (COM/2020/64). https://eur-lex.europa.eu/legal-content/en/TXT/?qid=1593079180383&uri=CELEX:52020DC0064 Accessed 14 November 2022.

Conboy, C. (2021, June 3). Response to report of killer robots in Libya. Stop Killer Robots. https://www.stopkillerrobots.org/news/response-to-report-of-killer-robots-in-libya/. Accessed November 11, 2022.

Condorelli, L. and Kress, C. (2010). The rules of attribution: General considerations. *In:* J. Crawford, et al. (Eds.), The Law of International Responsibility (Part III: The Sources of International Responsibility, Ch. 18). Oxford University Press.

Crawford, J., Pellet, A., Olleson, S. and Parlett, K. (2010). The Law of International Responsibility. *In:* Oxford University Press eBooks. https://doi.org/10.1093/law/9780199296972.001.0001.

Crootof, R. (2015). The killer robots are here: Legal and policy implications. Cardozo Law Review, 36, 1837. https://ssrn.com/abstract=2534567.

Crootof, R. (2016). War Torts: Accountability for Autonomous Weapons. University of Pennsylvania Law Review, 164. Retrieved from https://ssrn.com/abstract=2657680. Accessed January 1, 2019.

Crootof, R. (2018). Autonomous weapon systems and the limits of analogy. Harvard National Security Journal, 9, 51.

Cummings, M. (2019). Lethal autonomous weapons: Meaningful human control or meaningful human certification? [Opinion]. IEEE Technology and Society Magazine, 38, 20. https://ieeexplore.ieee.org/stamp/stamp.jsp?tp=&arnumber=8924577 Accessed 16 November 2022.

Daoust, I., Coupland, R. and Ishoey, R. (2002). New wars, new weapons? The obligation of states to assess the legality of means and methods of warfare. International Review of the Red Cross, 84(846), 345-363.

Del Monte, L. (2018). Genius Weapons: Artificial Intelligence, Autonomous Weaponry, and the Future of Warfare. Prometheus Books.

Davison, N. (2017). A legal perspective: Autonomous weapon systems under international humanitarian law. UNODA Occasional Papers No. 30 Perspectives on Lethal Autonomous Weapon Systems. Retrieved from https://www.icrc.org/en/download/file/65762/autonomous_weapon_systems_under_international_humanitarian_law.pdf Accessed 27 February 2019.

Dickinson, L. (2018). Lethal autonomous weapons systems: The overlooked importance of administrative accountability. *In:* E.T. Jensen & R. Alcala (Eds.), The Impact of Emerging Technologies on the Law of Armed Conflict. Oxford University Press (2018 Forthcoming). GWU Law School Public Law Research Paper (2018-42) Accessed 24 August 2023.

Docherty, B. (2012, November 19). Losing humanity: The case against killer robots. Human Rights Watch & Harvard Human Rights Clinic. https://www.hrw.org/report/2012/11/19/losing-humanity/case-against-killer-robots Accessed Mach 19, 2018.

Docherty, B. (2015). Mind the gap: The lack of accountability for killer robots. Human Rights Watch https://www.hrw.org/sites/default/files/reports/arms0415_ForUpload_0.pdf Accessed November 14, 2022.

Docherty, B. (2018). Heed the call: A moral and legal imperative to ban killer robots. Human Rights Watch. https://www.hrw.org/report/2018/08/21/heed-call/moral-and-legal-imperative-ban-killer-robots. Accessed January 22, 2022.

Docherty, B. (2019a). Statement to Convention on Conventional Weapons (CCW) Group of Governmental Experts on Lethal Autonomous Weapons Systems Agenda Item 5(a) regarding Challenges Posed to International Humanitarian Law. Human Rights Watch. https://docs-library.unoda.org/Convention_on_Certain_Conventional_Weapons_-_Group_of_Governmental_Experts_(2019)/IHL%2BGGE%2Bstatement-3%2B26%2B19-FINAL.pdf. Accessed November 14, 2022.

Docherty, B. (2019b). Elements of a Treaty on Fully Autonomous Weapons. Campaign to Stop Killer Robots. https://www.stopkillerrobots.org/wp-content/uploads/2020/03/Key-Elements-of-a-Treaty-on-Fully-Autonomous-Weapons.pdf Accessed November 13, 2022.

Docherty, B. (2022, March 16). Early signs of war crimes and human rights abuses committed by the Russian military during the full-scale invasion of Ukraine. Human Rights Watch. https://www.hrw.org/news/2022/03/16/early-signs-war-crimes-and-human-rights-abuses-committed-russian-military-during. Accessed November 16, 2022.

Dunlap Jr, C.J. (2016). Accountability and autonomous weapons: Much ado about nothing. Temp. Int'l & Comp. LJ, 30, 63.

Dutch Advisory Council on International Affairs (AIV) and Advisory Committee on Issues of Public International Law (CAVV). (2021). Advisory Report 119, CAVV-advisory report 38 https://www.advisorycouncilinternationalaffairs.nl/documents/publications/2021/12/03/autonomous-weapon-systems---summary-and-recommendations Accessed 17 November 2022.

Dutch Advisory Council on International Affairs (AIV) and Advisory Committee on Issues of Public International Law (CAVV). (2022). Autonomous Weapons Systems: The importance of regulation and investment.https://www.advisorycouncilinternationalaffairs.nl/documents/government-responses/2022/06/17/goverment-response-to-autonomous-weapon-systems-the-importance-of-regulation-and-investment Accessed August 22, 2023.

Ekelhof, M.A.C. (2017). Complications of a common language: Why it is so hard to talk about autonomous weapons. Journal of Conflict and Security Law, 22(3), 311.

Ekelhof, M. and Persi Paoli, G. (2020). Swarm Robotics: Technical and Operational Overview of the Next Generation of Autonomous Systems. UNIDIR. https://unidir.org/publication/swarm-robotics-technical-and-operational-overview-next-generation-autonomous-systems Accessed 14 November 2022.

European Commission. (2018). Communication from the Commission to the European Parliament, the European Council, the Council, the European Economic and Social Committee and the Committee of the Regions, 'Coordinated Plan on Artificial Intelligence' (COM(2018) 795 final). https://ec.europa.eu/digital-single-market/en/news/coordinated-plan-artificial-intelligence Accessed 14 November 2022.

European Commission. (2020). Report on the safety and liability implications of Artificial Intelligence, the Internet of Things, and robotics (COM/2020/64). https://eur-lex.europa.eu/legal-content/en/TXT/?qid=1593079180383&uri=CELEX:52020DC0064 Accessed 13 November 2022.

European Parliament and Council of Europe. (2016). Regulation 2016/679 - Protection of natural persons with regard to the processing of personal data and on the free movement of such data, and repealing Directive 95/46/EC (General Data Protection Regulation). (Publication Date: April 27, 2016).

European Parliament. (2018). Recommendation of 5 July 2018 to the Council on the 73rd session of the United Nations General Assembly (2018/2040(INI)) (P8_TA(2018)0312). https://www.europarl.europa.eu/doceo/document/TA-8-2018-0312_EN.pdf Accessed 16 November 2022.

European Union External Action Service. (2018, December 10). International Security and Lethal Autonomous Weapons. EEAS https://eeas.europa.eu/headquarters/headquarters-homepage/51679/international-security-and-lethal-autonomous-weapons_tr Accessed 16 November 2022.

Feige, I. (2019). What is AI Safety? Faculty AI blog. https://faculty.ai/blog/what-is-ai-safety/ Accessed 17 November 2022.

Finland, Estonia, France, Germany, and the Netherlands Delegations. (2019, May 17). Food for Thought Paper by Finland, Estonia, France, Germany, and the Netherlands. Digitalization and Artificial Intelligence in Defence. https://eu2019.fi/documents/11707387/12748699/Digitalization+and+AI+in+Defence.pdf/151e10fd-c004-c0ca-d86b-07c35b55b9cc/Digitalization+and+AI+in+Defence.pdf Accessed 17 November 2022.

Finland, France, Germany, the Netherlands, Norway, Spain, and Sweden Delegations. (2022, July 13). Working paper submitted by Finland, France, Germany, the Netherlands, Norway, Spain, and Sweden to the 2022 Chair of the Group of Governmental Experts (GGE) on emerging technologies in the area of lethal autonomous weapons systems (LAWS). https://documents.unoda.org/wp-content/uploads/2022/07/WP-LAWS_DE-ES-FI-FR-NL-NO-SE.pdf Accessed 17 November 2022.

Fletcher, L.E. (2016, February 1). A wolf in sheep's clothing? Transitional justice and the effacement of state accountability for international crimes. Fordham International Law Journal, 39, ssrn.com/abstract=2758803 Accessed 16 November 2022.

France (n.d.). French's Cour de cassation's website. https://www.courdecassation.fr/acces-rapide-judilibre Accessed 13 January 2022.

French, D. and Stephens, T. (2014). International Law Association (ILA) Study Group on Due Diligence in International Law First Report (Duncan French, Chair, and Tim Stephens, Rapporteur). (Publication Date: March 7, 2014).

French, D. and Stephens, T. (2016). International Law Association (ILA) Study Group on Due Diligence in International Law Second Report (Duncan French, Chair, and Tim Stephens, Rapporteur). (Publication Date: July 2016).

French delegation. (2018, August 28). Human-Machine Interaction in the Development, Deployment and Use of Emerging Technologies in the Area of Lethal Autonomous Weapons Systems (CCW/GGE.2/2018/WP.3). https://documents-dds-ny.un.org/

doc/UNDOC/GEN/G18/261/46/PDF/G1826146.pdf?OpenElement accessed 12 November 2022.

French delegation. (2020). Operationalization of the 11 guiding principles at national level – Comments by France. https://documents.unoda.org/wp-content/uploads/2020/07/20200610-France.pdf Accessed 16 November 2022.

French Ministère des Armées, Defense Ethics Committee. (2021, April 29). Opinion on the Integration of Autonomy into Lethal Weapons Systems. https://www.defense.gouv.fr/sites/default/files/ministere-armees/20210429_Comit%C3%A9%20d%27%C3%A9thique%20de%20la%20d%C3%A9fense%20-%20Avis%20int%C3%A9gration%20autonomie%20syst%C3%A8mes%20armes%20l%C3%A9taux%20-%20Version%20anglaise.pdf.pdf Accessed 26 May 2021.

Ford, C.M. (2017). Autonomous weapons and international law. South Carolina Law Review, 69(2), 413-476.

Future for Life Institute. (n.d.). Lethal Autonomous Weapons Systems. https://futureoflife.org/project/lethal-autonomous-weapons-systems/ Accessed 13 November 2022 and https://autonomousweapons.org/ Accessed 22 August 2023.

Geiß, R. (2014). Autonomous Weapons Systems: Risk Management and State Responsibility in Lethal Autonomous Weapons Systems Technology, Definition, Ethics, Law & Security. German Federal Foreign Office. https://www.auswaertiges-amt.de/blob/204830/5f26c2e0826db0d000072441fdeaa8ba/abruestung-laws-data.pdf Accessed November 17, 2022.

Geiß, R. (2015). Die Völkerrechtliche Dimension Autonomer Waffensysteme. Friedrich-Ebert-Stiftung, Internationale Politikanalyse. Retrieved from https://library.fes.de/pdf-files/id/ipa/11444-20150619.pdf. Accessed November 15, 2022.

Geiß, R. (2016). Autonomous Weapons Systems: Risk Management and State Responsibility. Third CCW meeting of experts on Lethal Autonomous Weapons Systems (LAWS) Geneva, 11-15 April 2016. https://docs-library.unoda.org/Convention_on_Certain_Conventional_Weapons_-_Informal_Meeting_of_Experts_(2016)/Geiss-CCW-Website.pdf. Accessed November 15, 2022.

Galvão Teles, P. (2021, August 3). The impact and influence of the Articles on State Responsibility on the work of the International Law Commission and beyond. EJIL Talk, https://www.ejiltalk.org/the-impact-and-influence-of-the-articles-on-state-responsibility-on-the-work-of-the-international-law-commission-and-beyond/. Accessed November 15, 2022.

Garcia, D. (2016). Future arms, technologies, and international law: Preventive security governance. European Journal of International Security, 1, 94.

Garcia, E. (2019, August 21). AI & Global Governance: When Autonomous Weapons Meet Diplomacy. United Nations University Centre for Policy Research. https://cpr.unu.edu/publications/articles/ai-global-governance-when-autonomous-weapons-meet-diplomacy.html Accessed 17 November 2022.

Gattini, A. (2014). Breach of International Obligations (SHARES Research Paper 31). Amsterdam Center for International Law, University of Amsterdam. http://www.sharesproject.nl/wp-content/uploads/2014/03/SHARES-RP-31-final.pdf. Accessed November 16, 2022.

Graefrath, B. (1984). Responsibility and Damages Caused: Relationship between Responsibility and Damages. *In:* Collected Courses of the Hague Academy of International Law (Vol. 185).

Group of Governmental Experts (GGE) of the high contracting parties to the convention on prohibitions or restrictions on the use of certain conventional weapons which may be deemed to be excessively injurious or to have indiscriminate effects (CCW). (2017, December 22). Report of the 2017 session of the Group of Governmental Experts on Emerging Technologies in the Area of Lethal Autonomous Weapons Systems (CCW/GGE.1/2017/3).

Group of Governmental Experts (GGE) of the high contracting parties to the convention on prohibitions or restrictions on the use of certain conventional weapons which may be deemed to be excessively injurious or to have indiscriminate effects (CCW). (2018, October 23). Report of the 2018 session of the Group of Governmental Experts on Emerging Technologies in the Area of Lethal Autonomous Weapons Systems 9–13 April 2018 and 27-31 August 2018 (CCW/GGE.1/2018/3). https://undocs.org/en/CCW/GGE.1/2018/3 Accessed 14 November 2022.

Group of Governmental Experts (GGE) of the high contracting parties to the convention on prohibitions or restrictions on the use of certain conventional weapons which may be deemed to be excessively injurious or to have indiscriminate effects (CCW). (2019a, September 25). Report of the 2019 session of the Group of Governmental Experts on Emerging Technologies in the Area of Lethal Autonomous Weapons Systems 25–29 March 2019 and 20–21 August 2019 (CCW/GGE.1/2019/3). https://undocs.org/en/CCW/GGE.1/2019/3 Accessed 14 November 2022.

Group of Governmental Experts (GGE) of the high contracting parties to the convention on prohibitions or restrictions on the use of certain conventional weapons which may be deemed to be excessively injurious or to have indiscriminate effects (CCW). (2019b, December 13). Final Report of the 2019 session of the Group of Governmental Experts on Emerging Technologies in the Area of Lethal Autonomous Weapons Systems 13–15 November 2019 (CCW/MSP/2019/9) Annex III Guiding Principles affirmed by the Group of Governmental Experts on Emerging Technologies in the Area of Lethal Autonomous Weapons Systems. https://documents-dds-ny.un.org/doc/UNDOC/GEN/G19/343/64/PDF/G1934364.pdf?OpenElement Accessed 16 November 2022.

Group of Governmental Experts (GGE) of the high contracting parties to the convention on prohibitions or restrictions on the use of certain conventional weapons which may be deemed to be excessively injurious or to have indiscriminate effects (CCW). (2021, April 219). Chairperson's Summary of the 2020 work of the Group of Governmental Experts on Emerging Technologies in the Area of Lethal Autonomous Weapons Systems (CCW/GGE.1/2020/WP.7).

Group of Governmental Experts (GGE) of the high contracting parties to the convention on prohibitions or restrictions on the use of certain conventional weapons which may be deemed to be excessively injurious or to have indiscriminate effects (CCW). (2022, February 22). Report of the 2021 session of the Group of Governmental Experts on Emerging Technologies in the Area of Lethal Autonomous Weapons Systems (CCW/GGE.1/2021/3).

Group of Governmental Experts (GGE) on emerging technologies in the area of lethal autonomous weapons systems (LAWS), Second session. (2022a, July 25-29). Transcripts automatic. https://indico.un.org/event/1001117/page/75-transcripts-automatic Accessed 15 November 2022.

Group of Governmental Experts (GGE) of the high contracting parties to the convention on prohibitions or restrictions on the use of certain conventional weapons which may be deemed to be excessively injurious or to have indiscriminate effects (CCW). (2022b, August 31). Report of the 2022 session of the Group of Governmental Experts on Emerging Technologies in the Area of Lethal Autonomous Weapons Systems 7–11

March, and 25–29 July 2022 (CCW/GGE.1/2022/2). https://unoda-documents-library.s3.amazonaws.com/Convention_on_Certain_Conventional_Weapons_-_Group_of_Governmental_Experts_(2022)/CCW_GGE1_2022_2_Final_Report_Advance_copy.pdf Accessed 15 November 2022.

Group of Governmental Experts (GGE) of the high contracting parties to the convention on prohibitions or restrictions on the use of certain conventional weapons which may be deemed to be excessively injurious or to have indiscriminate effects (CCW). (2023, May 24). Report of the 2023 session of the Group of Governmental Experts on Emerging Technologies in the Area of Lethal Autonomous Weapons Systems (CCW/GGE.1/2023/2).

Convention on prohibitions or restrictions on the use of certain conventional weapons which may be deemed to be excessively injurious or to have indiscriminate effects (2023, May 24).

Gutiérrez Espada, C. and Cervell Hortal, M.J. (2013). Sistemas de Armas Autónomas, Drones y Derecho Internacional. Revista del Instituto Español de Estudios Estratégicos, 2.

Guzman, Andrew T. and Meyer, Timothy. (2010, August 4). International Soft Law. The Journal of Legal Analysis, 2(1), Spring 2011, UC Berkeley Public Law Research Paper No. 1353444, Available at SSRN: https://ssrn.com/abstract=1353444 or http://dx.doi.org/10.2139/ssrn.1353444.

Hafner, G. (2004). Pros and cons ensuing from fragmentation of international law. Michigan Journal of International Law, 25(4). https://repository.law.umich.edu/mjil/vol25/iss4/2/ Accessed 17 November 2022.

Hammond, D.D. (2015). Autonomous weapons and the problem of state accountability. Chicago Journal of International Law, 15.

Heyns, C. (2013). Promotion and protection of all human rights, civil, political, economic, social and cultural rights, including the right to development: Report of the special rapporteur on extrajudicial, summary or arbitrary executions. Human Rights Council, Twenty-third session, Agenda Item 3, A/HRC/23/47. https://www.ohchr.org/sites/default/files/Documents/HRBodies/HRCouncil/RegularSession/Session23/A-HRC-23-47_en.pdf. Accessed February 18, 2019.

Heyns, C. (2014). Increasingly autonomous weapon systems: Accountability and responsibility. *In:* Autonomous Weapon Systems: Technical, Military, Legal and Humanitarian Aspects: Expert Meeting, ICRC, 26-28 March 2014. https://reliefweb.int/sites/reliefweb.int/files/resources/4221-002-autonomous-weapons-systems-full-report%20%281%29.pdf. Accessed November 14, 2022.

Heyns, C. (2017). Autonomous weapons in armed conflict and the right to a dignified life: An African perspective. South African Journal on Human Rights, 33(1), 46-71. DOI: 10.1080/02587203.2017.1303903.

High contracting parties to the convention on prohibitions or restrictions on the use of certain conventional weapons which may be deemed to be excessively injurious or to have indiscriminate effects (CCW). (2016, December 23). Final document of the 5th Review Conference (CCW/CONF.V/10).

High contracting parties to the convention on prohibitions or restrictions on the use of certain conventional weapons which may be deemed to be excessively injurious or to have indiscriminate effects (CCW). (2022, January 10). Final document of the 6th Review Conference (CCW/CONF.VI/11).

International Committee of the Red Cross (ICRC). (1987). Protocol Additional to the Geneva Conventions of 12 August 1949, and relating to the Protection of Victims of

International Armed Conflicts (Protocol I), 8 June 1977. Commentary of 1987 Article 91 – Responsibility https://ihl-databases.icrc.org/applic/ihl/ihl.nsf/Comment.xsp?action=openDocument&documentId=1066AF25ED669409C12563CD00438071 Accessed 17 November 2022.

International Committee of the Red Cross (ICRC). (2006, December). A guide to the legal review of new weapons, means and methods of warfare: Measures to implement Article 36 of additional protocol I of 1977 Geneva, Volume 88 Number 864. https://www.icrc.org/en/doc/assets/files/other/irrc_864_icrc_geneva.pdf Accessed 16 November 2022.

International Committee of the Red Cross (ICRC). (2015, April 15). Autonomous weapon systems: Is it morally acceptable for a machine to make life and death decisions? Statement of the ICRC. Convention on Certain Conventional Weapons (CCW), Meeting of Experts on Lethal Autonomous Weapons Systems (LAWS), 13-17 April 2015, Geneva. https://www.icrc.org/en/document/lethal-autonomous-weapons-systems-laws Accessed 13 November 2022.

International Committee of the Red Cross (ICRC). (2016, April 11). Views of the ICRC on autonomous weapon systems, paper submitted to the CCW Experts. Meeting of Experts on Lethal Autonomous Weapons Systems (LAWS), 11-15 April 2016, Geneva. https://www.icrc.org/en/document/views-icrc-autonomous-weapon-system Accessed 13 November 2022.

International Committee of the Red Cross (ICRC). (2018, April 3). Ethics and autonomous weapon systems: An ethical basis for human control? Geneva. https://www.icrc.org/en/document/ethics-and-autonomous-weapon-systems-ethical-basis-human-control Accessed 15 November 2022.

International Committee of the Red Cross (ICRC). (2019a, June) Artificial intelligence and machine learning in armed conflict: A human-centred approach https://www.icrc.org/en/document/artificial-intelligence-and-machine-learning-armed-conflict-human-centred-approach Accessed 13 November 2022.

International Committee of the Red Cross (ICRC). (2019b, August), Autonomy, artificial intelligence and robotics: Technical aspects of human control. https://www.icrc.org/en/download/file/102852/autonomy_artificial_intelligence_and_robotics.pdf Accessed 13 November 2022.

International Committee of the Red Cross (ICRC). (2019c). Statement of the International Committee of the Red Cross (ICRC) under agenda item 5(e) Possible options for addressing the humanitarian and international security challenges posed by emerging technologies in the area of lethal autonomous weapon systems in the context of the objectives and purposes of the Convention without prejudicing policy outcomes and taking into account past, present and future proposals. Meeting of Group of Governmental Experts on Lethal Autonomous Weapons Systems (LAWS) 25-29 March 2019, Geneva. https://docs-library.unoda.org/Convention_on_Certain_Conventional_Weapons_-_Group_of_Governmental_Experts_(2019)/CCW%2BGGE%2BLAWS%2BICRC%2Bstatement%2Bagenda%2Bitem%2B5e%2B27%2B03%2B2019.pdf Accessed 16 November 2022.

International Committee of the Red Cross (ICRC). (2020a). Position on autonomous weapon systems: ICRC position and background paper. International Review of the Red Cross, 102, 1335. https://international-review.icrc.org/sites/default/files/reviews-pdf/2022-01/icrc-position-on-autonomous-weapon-systems-icrc-position-and-background-paper-915.pdf Accessed 16 November 2022.

International Committee of the Red Cross (ICRC). (2020b). Statement of the International Committee of the Red Cross Convention on Certain Conventional Weapons (CCW) - Group of Governmental Experts on Lethal Autonomous Weapons Systems (LAWS) 21-25 September 2020. Geneva. https://documents.unoda.org/wp-content/uploads/2020/09/20200921-ICRC-General-statement-CCW-GGE-LAWS-Sep-2020.pdf Accessed 13 November 2022.

International Committee of the Red Cross (ICRC). (2021). Statement of the International Committee of the Red Cross Convention on Certain Conventional Weapons (CCW) Group of Governmental Experts on Lethal Autonomous Weapons Systems (LAWS) 3-13 August 2021. Geneva. https://documents.unoda.org/wp-content/uploads/2021/08/ICRC-statement-CCW-GGE-LAWS-August-2021_en.pdf Accessed 13 November 2022.

International Committee of the Red Cross (ICRC). (n.d.a). Rule 1. The Principle of Distinction between Civilians and Combatants. *In:* IHL Database Customary IHL. [URL] Accessed 15 November 2022. https://ihl-databases.icrc.org/customary-ihl/eng/docs/v1_rul_rule1 Accessed 15 November 2022.

International Committee of the Red Cross (ICRC). (n.d.b). Rule 17. Choice of Means and Methods of Warfare. *In:* IHL Database Customary IHL. [URL] Accessed 15 November 2022. https://ihl-databases.icrc.org/customary-ihl/eng/docs/v1_cha_chapter5_rule17 and https://ihl-databases.icrc.org/en/customary-ihl/v2/rule17 Accessed 15 November 2022.

International Committee of the Red Cross (ICRC). (n.d.c). Rule 22. Principle of Precautions against the Effects of Attacks. *In:* IHL Database Customary IHL. [URL] Accessed 15 November 2022. https://ihl-databases.icrc.org/customary-ihl/eng/docs/v1_rul_rule22 Accessed 15 November 2022.

International Committee of the Red Cross (ICRC). (n.d.d). Rule 144. Ensuring Respect for International Humanitarian Law Erga Omnes. *In:* IHL Database Customary IHL. [URL] Accessed 15 November 2022. https://ihl-databases.icrc.org/customary-ihl/eng/docs/v1_cha_chapter41_rule144 Accessed 15 November 2022.

International Committee of the Red Cross (ICRC). (n.d.e). Rule 149. Responsibility for violations of International Humanitarian Law. *In:* IHL Database Customary IHL. [URL] Accessed 15 November 2022. https://ihl-databases.icrc.org/customary-ihl/eng/docs/v1_rul_rule149#refFn_EC46E6AE_00001 Accessed 15 November 2022.

IEEE. (2019, June 24). IEEE Position Statement Artificial Intelligence. Approved by the IEEE Board of Directors. [PDF document]. https://globalpolicy.ieee.org/wp-content/uploads/2019/06/IEEE18029.pdf Accessed 13 November 2022.

IEEE. (2021a, September 23). IEEE Ontological Standard 7007 for Ethically Driven Robotics and Automation Systems Developed (Approved 23 September 2021 IEEE SA Standards Board). https://ieeexplore.ieee.org/document/9611206/ Accessed 16 November 2022.

IEEE. (2021b). IEEE Standard 7001 for Transparency of Autonomous Systems. https://standards.ieee.org/ieee/7001/6929/ Accessed 15 December 2022.

International Panel on the Regulation of Autonomous Weapons (iPRAW). (2020, September). Commentary on the Guiding Principles. https://documents.unoda.org/wp-content/uploads/2020/09/iPRAW_Commentary_GuidingPrinciples.pdf Accessed 10 November 2022.

Israel Permanent Mission to the UN Geneva. (2016, April 11). Statement on Lethal Autonomous Weapons Systems (LAWS) by Mrs. Maya Yaron, Counsellor Deputy

Permanent Representative to the Conference on Disarmament. https://docs-library.unoda.org/Convention_on_Certain_Conventional_Weapons_-_Informal_Meeting_of_Experts_(2016)/2016_LAWS_MX_GeneralDebate_Statements_Israel.pdf Accessed November 16, 2022.

Israel Permanent Mission to the UN Geneva. (2018a, April 11). Statement by Mrs. Maya Yaron, Minister – Counsellor Deputy Permanent Representative to the Conference on Disarmament. The Convention on Certain Conventional Weapons (CCW) GGE on Lethal Autonomous Weapons Systems (LAWS) Human Machine Interface. https://docs-library.unoda.org/Convention_on_Certain_Conventional_Weapons_-_Group_of_Governmental_Experts_(2018)/2018_LAWS6b_Israel.pdf Accessed November 12, 2022.

Israel Permanent Mission to the UN Geneva. (2018b, August 29). Statement by Mr. Ofer Moreno Director, Arms Control Department, Strategic Affairs Division, Ministry of Foreign Affairs. https://docs-library.unoda.org/Convention_on_Certain_Conventional_Weapons_-Groupof_Governmental_Experts_(2018)/2018_GGE%2BLAWS%2B2_6d_Israel.pdf Accessed November 16, 2022.

Israel Permanent Mission to the UN Geneva. (2019a, March 27). Statement by Mrs. Maya Yaron, Minister – Counsellor Deputy Permanent Representative to the Conference on Disarmament. Group of Governmental Experts on Emerging Technologies in the Area of Lethal Autonomous Weapons Systems. https://docs-library.unoda.org/Convention_on_Certain_Conventional_Weapons_-_Group_of_Governmental_Experts_(2019)/israel%2Bintervention%2B-%2BGGE%2BLAWS%2B-%2Bpossible%2Bpolicy%2Boptions%2B27.3.2019.pdf Accessed November 16, 2022.

Israel Permanent Mission to the UN Geneva. (2019b, March 26). Statement by Mr. Asaf Segev, Arms Control Department, Ministry of Foreign Affairs to the Group of Governmental Experts on Emerging Technologies in the Area of Lethal Autonomous Weapons Systems. https://docs-library.unoda.org/Convention_on_Certain_Conventional_Weapons_-_Group_of_Governmental_Experts_(2019)/CCW%2B-%2BGGE%2BLAWS%2B-%2BIsrael%2Bintervention%2B-%2BHuman%2BMachine%2BInterface%2B-%2B26%2BMarch%2B2019.pdf Accessed November 16, 2022.

Israel Permanent Mission to the UN Geneva. (2020, August 31). Israel Considerations on the Operationalization of the Eleven Guiding Principles Adopted by the Group of Governmental Experts. https://documents.unoda.org/wp-content/uploads/2020/09/20200831-Israel.pdf Accessed November 16, 2022.

Israel. (n.d.). Israeli Supreme Court Website. https://supreme.court.gov.il/sites/en/Pages/fullsearch.aspx Accessed January 13, 2022.

Jevglevskaja, N. (2018). Weapons review obligation under customary international law. International Law Studies, 186. https://digital-commons.usnwc.edu/cgi/viewcontent.cgi?article=1724&context=ils Accessed 16 November 2022.

Kallenborn, Z. (2022, March 15). Russia may have used a killer robot in Ukraine. The Bulletin. https://thebulletin.org/2022/03/russia-may-have-used-a-killer-robot-in-ukraine-now-what/ Accessed 27 August 2023.

Kalmanovitz, P. (2016). Judgement, liability and the risk of riskless warfare. *In:* Bhuta, N., Beck, S., Geiβ, R., Liu, H. & Kreβ, C. (Eds.), Autonomous Weapons Systems: Law, Ethics, Policy. Cambridge University Press.

Kania, E. (2018, April 17). China's strategic ambiguity and shifting approach to lethal autonomous weapons systems. Lawfare. https://www.lawfaremedia.org/article/

chinas-strategic-ambiguity-and-shifting-approach-lethal-autonomous-weapons-systems. Accessed September 19, 2023.

Kant, I. (1998). Groundwork of the Metaphysics of Morals. (Trans. & Ed., M. Gregor). Cambridge University Press.

Koivurova, T. (2010). Due Diligence. Max Planck Encyclopedias of Public International Law [MPEPIL]. Published under the auspices of the Max Planck Institute for Comparative Public Law and International Law under the direction of Professor Anne Peters (2021–) and Professor Rüdiger Wolfrum (2004–2020). https://opil.ouplaw.com/view/10.1093/law:epil/9780199231690/law-9780199231690-e1034 Accessed 17 November 2022.

Kolb, R. (2017). The International Law of State Responsibility. Edward Elgar Publishing.

Konaev, M. and Bendett, S. (2019, July 31). Russian AI-enabled combat: Coming to a city near you? Warontherocks. https://warontherocks.com/2019/07/RUSSIAN-AI-ENABLED-COMBAT-COMING-TO-A-CITY-NEAR-YOU/ Accessed 8 October 2022.

Korea. (n.d.). Korean Supreme Court. https://eng.scourt.go.kr/eng/main/Main.work Accessed January 13, 2022.

Koskenniemi, M. and Leino, P. (2002). Fragmentation of international law? Postmodern anxieties. Leiden Journal of International Law, 15(3), 553-579. doi:10.1017/S0922156502000262.

Law, J. and Martin, E.A. (2009). A dictionary of law. *In:* Oxford University Press eBooks. 7ed. https://doi.org/10.1093/acref/9780199551248.001.0001.

Lieblich, E. and Benvenisti, E. (2016). The obligation to exercise discretion in warfare: Why autonomous weapons systems are unlawful. *In:* N. Bhuta & others (Eds.), Autonomous Weapons Systems: Law, Ethics, Policy (1st ed., pp. 250). Cambridge University Press.

Long, R.E. (2021, August 16). Artificial Intelligence Liability: The Rules Are Changing. Retrieved from https://blogs.lse.ac.uk/businessreview/2021/08/16/artificial-intelligence-liability-the-rules-are-changing/. Accessed November 16, 2022.

Mačák, K. (2021, August 6). Strengthening the rule of law in time of war: An IHL perspective on the present and future of the Articles on State Responsibility. EJIL Talk. https://www.ejiltalk.org/strengthening-the-rule-of-law-in-time-of-war-an-ihl-perspective-on-the-present-and-future-of-the-articles-on-state-responsibility/ Accessed September 19, 2023.

Mann, J.-K. (2019). Autonomous weapons systems and the liability gap, Part Two: Civil Liability and State Responsibility. Rethinking SLIC. https://rethinkingslic.org/blog/53-autonomous-weapons-systems-and-the-liability-gap-part-two-civil-liability-and-state-responsibility Accessed 16 November 2022.

Margulies, P. (2016). Making autonomous weapons accountable: Command responsibility for computer-guided lethal force in armed conflicts. *In:* J.D. Ohlin (Ed.), Research Handbook on Remote Warfare. Edward Elgar Press. Forthcoming. Roger Williams Univ. Legal Studies Paper No. 166. https://ssrn.com/abstract=2734900 Accessed 14 November 2022.

Margulies, P. (2018). The other side of autonomous weapons: Using artificial intelligence to enhance IHL compliance. *In:* E.T. Jensen (Ed.), The Impact of Emerging Technologies on the Law of Armed Conflict. Oxford University Press. Forthcoming. Lieber Institute for Law and Land Warfare, U.S. Military Academy at West Point. Roger Williams Univ. Legal Studies Paper No. 182. https://ssrn.com/abstract=3194713 Accessed 16 November 2022.

Marino, D. and Tamburrini, G. (2006). Learning robots and human responsibility. The International Review of Information Ethics, 6, 46.

Mashaw, J.L. (2006). Accountability and institutional design: Some thoughts on the grammar of governance. *In:* M. Dowdle (Ed.), Public Accountability: Designs, Dilemmas and Experiences (pp. 115–156). Cambridge University Press.

Matthias, A. (2004). The responsibility gap: Ascribing responsibility for the actions of learning automata. Ethics and Information Technology, 6(3), 175–183. https://doi.org/10.1007/s10676-004-3422-1.

Mauri, D. (2022). Autonomous Weapons Systems and the Protection of the Human Person: An International Law Analysis. Edward Elgar Publishing.

McDougall, C. (2019). Autonomous weapon systems and accountability: Putting the cart before the horse. Melbourne Journal of International Law, 20(1), 58–87.

Meron, T. and Denison, C. (2000). Marco Sassòli and Antoine Bouvier, in cooperation with Laura M. Olson, Nicolas A. Dupic and Lina Milner How Does Law Protect in War? Cases, Documents, and Teaching Materials on Contemporary Practice in International Humanitarian Law International Committee of the Red Cross, Geneva, 1999, 1,493 pages. International Review of the Red Cross. https://doi.org/10.1017/s1560775500184779.

Montavon, G., Samek, W. and Müller, K.-R. (2018). Methods for interpreting and understanding deep neural networks. Digital Signal Processing, 73, 1.

Moyes, R. (2013, April). Killer robots: UK government policy on fully autonomous weapons. Article 36. Retrieved from http://www.article36.org/wp-content/uploads/2013/04/Policy_Paper1.pdf. Accessed August 25, 2023.

Moyes, R. (2019). Target profiles, discussion paper regarding 'autonomy' and weapons systems. Article 36. https://article36.org/wp-content/uploads/2019/08/Target-profiles.pdf. Accessed November 13, 2022.

Moynihan, H. (2018). Aiding and assisting: The mental element under Article 16 of the International Law Commission's Articles on State Responsibility. International and Comparative Law Quarterly, 67, 455.

Murdock, J. (2020). Tesla Faces Lawsuit after Model X on Autopilot with 'Dozing Driver' Blamed for Fatal Crash. Newsweek. https://www.newsweek.com/tesla-lawsuit-model-x-autopilot-fatal-crash-japan-yoshihiro-umeda-1501114 Accessed 16 November 2022.

Myers, J. (2018). Speech at Carnegie Council that opened the presentation of Paul Scharre on his book Army of None. Carnegie Council. https://www.carnegiecouncil.org/studio/multimedia/20180501-army-of-none-autonomous-weapons-future-of-war-paul-scharre Accessed 1 November 2019.

NATO Terminology Database. (2020). Autonomous, Record 40489. NATO Glossaries: AAP-06 TTF: 2019-0294 (Approval date: 13 May 2020) https://nso.nato.int/natoterm/Web.mvc Accessed 13 November 2022.

Noble, S.W. (2006). What is a support vector machine? Nature Biotechnology, 24, 1565.

Oxford Learner's Dictionaries. (2022a). Select. https://www.oxfordlearnersdictionaries.com/us/definition/english/select_1?q=select Accessed 12 September 2022 .

Oxford Learner's Dictionaries. (2022b). Target. https://www.oxfordlearnersdictionaries.com/us/definition/english/target_1?q=target Accessed 12 September 2022.

Oxford Public International Law. (n.d.). Oxford International Law and Domestic Courts Database. https://opil.ouplaw.com/search?ct=6f151ede-c53f-4613-8bea-f62a076a1355 Accessed 12 September 2022.

Oxford Reference. (n.d.) Weapons. https://www.oxfordreference.com/view/10.1093/oi/authority.20110803121436977 accessed12 Sep. 2022.

Pellet, A. (2000, July 25). Report of the Working Group on Long-term Program of Work. United Nations Digital Library System. ILC(LII)/WG/LT/L.1/Add.1 https://digitallibrary.un.org/record/419775?ln=en Accessed November 17, 2022.

Pemmaraju, S.R. (2000). Third report on international liability for injurious consequences arising out of acts not prohibited by international law (prevention of transboundary damage from hazardous activities) [Agenda item 4]. Document A/CN.4/510. United Nations. https://legal.un.org/ilc/documentation/english/a_cn4_510.pdf .

People's Republic of China. (2017). China's 2017 State Council New Generation Artificial Intelligence Development Plan. Translated by Webster, Creemers, Kania & Triolo https://www.newamerica.org/cybersecurity-initiative/digichina/blog/full-translation-chinas-new-generation-artificial-intelligence-development-plan-2017/. Accessed November 17, 2022.

People's Republic of China. (2018, April 11). Group of governmental experts of the high contracting parties to the convention on prohibitions or restrictions on the use of certain conventional weapons which may be deemed to be excessively injurious or to have indiscriminate effects, Geneva, 9–13 April 2018 (first week) Item 6 of the provisional agenda. Other matters Position Paper Submitted by China (CCW/GGE.1/2018/WP.7). https://www.reachingcriticalwill.org/images/documents/Disarmament-fora/ccw/2018/gge/documents/GGE.1-WP7.pdf. Accessed November 16, 2022.

People's Republic of China. (2020, January 9). People's Court of the People's Republic of China Supreme Court website in English. http://english.court.gov.cn/index.html. Accessed January 13, 2022.

People's Republic of China Delegation. (2022, July). Working Paper of the People's Republic of China on Lethal Autonomous Weapons Systems (Unofficial Translation). https://documents.unoda.org/wp-content/uploads/2022/07/Working-Paper-of-the-Peoples-Republic-of-China-on-Lethal-Autonomous-Weapons-Systems%EF%BC%88English%EF%BC%89.pdf. Accessed November 11, 2022.

Perrigo, B. (2018, April 9). A global arms race for killer robots is transforming the battlefield. Time. https://time.com/5230567/killer-robots/. Accessed August 10, 2022.

Peters, A. (2017). The refinement of international law: From fragmentation to regime interaction and politicization. Icon-international Journal of Constitutional Law, 15(3), 671–704. https://doi.org/10.1093/icon/mox056.

Petman, J.M. (2017). Autonomous Weapons Systems and International Humanitarian Law: 'Out of the Loop'? (Research reports). The Eric Castren Institute of International Law and Human Rights. https://researchportal.helsinki.fi/en/publications/autonomous-weapons-systems-and-international-humanitarian-law-out Accessed 13 November 2022.

Pilloud, C. and others (Eds.). (1987). Commentary on the Additional Protocols of 8 June 1977 to the Geneva Conventions of 12 August 1949. Martinus Nijhoff Publishers.

Pourcel, E. (2018). Dronisation et Robotisation Intelligentes Des Armées (DRIA): De La Dynamique Conflictuelle et Opérationnelle Mixte Homme-Machine... à La Dynamique Conflictuelle et Opérationnelle Machine IA-Machine IA? L'Harmattan.

Prestes, E. and others. (2013). Towards a core ontology for robotics and automation. Robotics and Autonomous Systems, 61, 1193. https://www.sciencedirect.com/science/article/abs/pii/S0921889013000596 Accessed 16 November 2022.

Prestes, E. and others. (2021). The first global ontological standard for ethically driven robotics and automation systems [Standards]. IEEE Robotics & Automation

Magazine, 28, 120. https://research-portal.uws.ac.uk/en/publications/the-first-global-ontological-standard-for-ethically-driven-roboti Accessed 16 November 2022.

Program on Humanitarian Policy and Conflict Research at Harvard University (HPCR), International Humanitarian Law Research Initiative. (2010). HPCR Manual on International Law Applicable to Air and Missile Warfare. https://reliefweb.int/sites/reliefweb.int/files/resources/8B2E79FC145BFB3D492576E00021ED34-HPCR-may2009.pdf Accessed September 19, 2023.

Quell, M. (2022, January 26). European Rights Court Hears Case Against Russia Over MH17 Crash. Courthouse News Service. https://www.courthousenews.com/european-rights-court-hears-case-against-russia-over-mh17-crash/ Accessed 14 November 2022.

Republic of Korea Delegation. (2021, August 4). Statement made on Aug. 4, 2021, at the Group of Governmental Experts on Emerging Technologies in the Area of Lethal Autonomous Weapons Systems. Audio and transcripts available at https://indico.un.org/event/35882/page/0. Accessed September 1, 2022.

Republic of Korea Permanent Mission Convention on Certain Conventional Weapons (CCW). (2014, May 13). Statement by Ambassador AHN Youngjip at the Meeting of Experts on Lethal Autonomous Weapons Systems in Geneva. Retrieved from https://overseas.mofa.go.kr/ch-geneva-en/brd/m_8813/view.do?seq=713230&page=1. Accessed November 12, 2022.

Reynolds, E. (2018, June 1). The agony of Sophia, the world's first robot citizen condemned to a lifeless career in marketing. Wired. https://www.wired.co.uk/article/sophia-robot-citizen-womens-rights-detriot-become-human-hanson-robotics Accessed 17 November 2022.

Russian Federation Delegation. (2018, April 4). Russia's Approaches to the Elaboration of a Working Definition and Basic Functions of Lethal Autonomous Weapons Systems in the Context of the Purposes and Objectives of the Convention (CCW/GGE.1/2018/WP.6). Presented at the Group of Governmental Experts on Lethal Autonomous Weapons Systems. [Version in English available at] https://reachingcriticalwill.org/images/documents/Disarmament-fora/ccw/2018/gge/documents/GGE.1-WP6-English.pdf. Accessed November 12, 2022.

Russian Federation Delegation. (2020). Working Paper of the Russian Federation, National Implementation of the Guiding Principles on Emerging Technologies in the Area of Lethal Autonomous Weapons Systems [Unofficial translation]. https://documents.unoda.org/wp-content/uploads/2020/09/Ru-Commentaries-on-GGE-on-LAWS-guiding-principles1.pdf. Accessed November 16, 2022.

Russian Federation Delegation. (2022, July 18). Working Paper Application of International Law to Lethal Autonomous Weapons Systems (LAWS). Submitted to Group of Governmental Experts on Lethal Autonomous Weapons Systems. https://documents.unoda.org/wp-content/uploads/2022/07/WP-Russian-Federation_EN.pdf. Accessed November 12, 2022.

Saliba, A. and Barbosa, L. (2023). Verbete Responsabilidade Internacional. *In:* D. Moura, F. Loureiro Bastos, & others (Eds.), Enciclopédia Luso-Brasileira de Direito Internacional. Editora Dom Quixote.

Sanders, L. and Copeland, D. (2020, November 27). Developing an approach to the legal review of autonomous weapon systems. ILA Reporter. http://ilareporter.org.au/2020/11/developing-an-approach-to-the-legal-review-of-autonomous-weapon-systems-lauren-sanders-and-damian-copeland/ Accessed 16 November 2022.

Sartor, G. and Omcini, A. (2016). The autonomy of technological systems and responsibilities for their use. *In:* N. Bhuta & others (Eds.), Autonomous Weapons Systems: Law, Ethics, Policy (1st ed., Chapter Title). Cambridge University Press.

Sassòli, M. (2002). State Responsibility for Violations of International Humanitarian Law. RICR Juin IRRC June 2002, 84(846).

Sassòli, M. (2014). Autonomous weapons and international humanitarian law: Advantages, open technical questions and legal issues to be clarified. International Law Studies, 90, 1. https://digital-commons.usnwc.edu/cgi/viewcontent.cgi?article=1017&context=ils Accessed 16 November 2022.

Sassòli, M. and Nagler, P. (2019). International Humanitarian Law: Rules, Controversies, and Solutions to Problems Arising in Warfare. Edward Elgar Publishing.

Saxon, D. (2016). A human touch: Autonomous weapons, DoD Directive 3000.09 and the interpretation of 'appropriate levels of human judgment over the use of force.' *In:* N. Bhuta & others (Eds.), Autonomous Weapons Systems: Law, Ethics, Policy (1st ed., pp. 185-208). Cambridge University Press.

Scharre, P. (2016, February). Autonomous Weapons and Operational Risk. Ethical Autonomy Project (Center for a New American Security), https://s3.amazonaws.com/files.cnas.org/documents/CNAS_Autonomous-weapons-operational-risk.pdf. Accessed November 13, 2022.

Scharre, P. (2018). Army of None: Autonomous Weapons and the Future of War (1st ed.). Norton & Company.

Schmitt, M. (2013), Tallinn Manual on the International Law Applicable to Cyber Warfare. Cambridge: Cambridge University Press.

Schmitt, M. (2015, August 10). Regulating autonomous weapons might be smarter than banning them. Just Security. https://www.justsecurity.org/25333/regulating-autonomous-weapons-smarter-banning/ Accessed 14 November 2022.

Schmitt, M. (2017). Tallinn Manual 2.0 on the International Law Applicable to Cyber Operations (2nd ed.). Cambridge: Cambridge University Press. doi:10.1017/9781316822524.

Secrétariat Général de la Défense et de la Sécurité Nationale. (2017). Chocs Futurs: Étude prospective à l'horizon 2030: impacts des transformations et ruptures technologiques sur notre environnement stratégique et de sécurité. https://www.frstrategie.org/sites/default/files/documents/publications/autres/2017/2017-sgdsn-chocs-futurs.pdf Accessed 10 March 2022.

SHARES Project (n.d.) Research Project on Shared Responsibility in International Law, developed by the Amsterdam Center for International Law, University of Amsterdam. (n.d.). http://www.sharesproject.nl/ Accessed 16 November 2022.

Slijper, F., Beck, A. and Kayser, D. (2019). State of AI Artificial Intelligence, the Military and Increasingly Autonomous Weapons. Susan Clark Translations, PAX for Peace Institute. https://paxforpeace.nl/media/download/state-of-artificial-intelligence--pax-report.pdf accessed 1 November 2022.

Soni, A. and Dominic, E. (2020). Legal and Policy Implications of Autonomous Weapons Systems. Center of Internet and Society. https://cis-india.org/internet-governance/blog/legal-and-policy-implications-of-autonomous-weapons-systems. Accessed November 6, 2022.

Sparrow, R. (2007). Killer robots. Journal of Applied Philosophy, 24(1), 62–77.

Statista. (2021). Countries with the highest military spending worldwide in 2021 (in billion U.S. dollars). https://www.statista.com/statistics/262742/countries-with-the-highest-military-spending/ Accessed 30 October 2022.

STM. (n.d.). KARGU® Combat Proven Rotary Wing Loitering Munition System. https://www.stm.com.tr/en/kargu-autonomous-tactical-multi-rotor-attack-uav Accessed 22 October 2022.

Stoltenberg, J. (2019). Speech by NATO Secretary General Stoltenberg at the High-level NATO Conference on Arms Control and Disarmament. NATO. https://www.nato.int/cps/en/natohq/opinions_169930.htm?selectedLocale=en Accessed 2 September 2020.

Stürchler, N. and Siegrist, M. (2017). A "Compliance-Based" Approach to Autonomous Weapon Systems. EJIL Talk. https://www.ejiltalk.org/a-compliance-based-approach-to-autonomous-weapon-systems/ Accessed 15 November 2022.

Sunstein, C.R. and Vermeule, A. (2006). Is Capital Punishment Morally Required? Acts, Omissions, and Life-Life Tradeoffs. Stanford Law Review, 58(3), 703-750.

Switzerland Delegation. (2017). A "compliance-based" approach to Autonomous Weapon Systems Working Paper submitted by Switzerland (Document Number: CCW/GGE.1/2017/WP.9). https://www.reachingcriticalwill.org/images/documents/Disarmament-fora/ccw/2017/gge/documents/WP9.pdf Accessed 16 November 2022.

Taddeo, M. and Blanchard, A. (2023). A comparative analysis of the definitions of autonomous weapons. *In:* Digital Ethics Lab Yearbook (pp. 57–79). https://doi.org/10.1007/978-3-031-28678-0_6.

Tamburrini, G. (2016). On banning autonomous weapons systems: From deontological to wide consequentialist reasons. *In:* N. Bhuta & others (Eds.), Autonomous Weapons Systems: Law, Ethics, Policy (1st ed., pp. 122-142). Cambridge University Press.

Tams, C. and Paddeu, F. (2021, August 9). Dithering, Trickling Down, and Encoding: Concluding Thoughts on the 'ILC Articles at 20' Symposium. EJIL Talk. https://www.ejiltalk.org/dithering-trickling-down-and-encoding-concluding-thoughts-on-the-ilc-articles-at-20-symposium/. Accessed November 15, 2022.

Techopedia. (2016). Support Vector Machine. [Dictionary > Data Management > Support Vector Machine]. Editor: Margaret Rouse. https://www.techopedia.com/definition/30364/support-vector-machine-svm, . Editor: Margaret Rouse. Last updated: September 14, 2016. Accessed August 25, 2023.

Techopedia. (2023). Deep Neural Network. [Dictionary > Artificial Intelligence > Deep Neural Network]. Editor: Margaret Rouse. https://www.techopedia.com/definition/32902/deep-neural-network. Last updated: July 21, 2023. Accessed August 25, 2023.

United Kingdom (UK). (2017). Joint Doctrine Publication 0-30.2, Unmanned Aircraft Systems (Published 12 September 2017 - Last updated 15 January 2018). https://www.gov.uk/government/publications/unmanned-aircraft-systems-jdp-0-302 Accessed 12 November 2022.

United Kingdom (UK) Centre for Data Ethics and Innovation. (2023). Algorithmic Transparency Standard: Guidance for Public Sector Organizations. https://github.com/co-cddo/algorithmic-transparency-standard/blob/main/Guidance%20for%20Public%20Sector%20Organisations%E2%80%99%20Use%20of%20the%20Algorithmic%20Transparency%20Standard%20v1.1.pdf Accessed September 19, 2023.

United Kingdom (UK) Delegation. (2018, August 27-31). Convention on Certain Conventional Weapons Group of Government Experts on emerging technologies in the area of Lethal Autonomous Weapons Systems, Item 6 of the provisional agenda: Other matters. Human Machine Touchpoints: The United Kingdom's perspective on human control over weapon development and targeting cycles. CCW/GGE.2/2018/

WP.1 Submitted by the United Kingdom (8 August 2018) https://reachingcriticalwill.org/images/documents/Disarmament-fora/ccw/2018/gge/documents/GGE.2-WP1.pdf Accessed 12 November 2022.

United Kingdom (UK) Delegation. (2019, March 25-29). Convention on Certain Conventional Weapons Group of Government Experts on emerging technologies in the area of Lethal Autonomous Weapons Systems, Agenda item 5(b): Characterization of the systems under consideration in order to promote a common understanding on concepts and characteristics relevant to the objectives and purposes of the Convention. https://unoda-documents-library.s3.amazonaws.com/Convention_on_Certain_Conventional_Weapons_-_Group_of_Governmental_Experts_(2019)/20190318-5%28b%29_Characterisation_Statement.pdf Accessed 12 November 2022.

United Kingdom (UK) Delegation. (2020). Convention on Certain Conventional Weapons Group of Government Experts on emerging technologies in the area of Lethal Autonomous Weapons Systems, UK Commentary on the Operationalization of the LAWS Guiding Principles. https://documents.unoda.org/wp-content/uploads/2020/09/20200901-United-Kingdom.pdf Accessed 12 November 2022.

United Kingdom (UK) Delegation. (2022). Convention on Certain Conventional Weapons Group of Government Experts on emerging technologies in the area of Lethal Autonomous Weapons Systems, Proposal for a GGE document on the application of International Humanitarian Law to Emerging Technologies in the Area of Lethal Autonomous Weapons Systems - Proposed by the United Kingdom. https://documents.unoda.org/wp-content/uploads/2022/05/03032022-UK-Proposal-for-Mar-2022-LAWS-GGE.docx Accessed 12 November 2022.

United Kingdom (UK) Ministry of Defence (MoD). (2022, June). Defence Artificial Intelligence Strategy 2022. [PDF]. UK.gov. https://assets.publishing.service.gov.uk/government/uploads/system/uploads/attachment_data/file/1082416/Defence_Artificial_Intelligence_Strategy.pdf Accessed 12 November 2022.

United Nations (UN) Committee on the Rights of the Child (CRC). (2003, November 27). General comment no. 5: General measures of implementation of the Convention on the Rights of the Child, CRC/GC/2003/5.

United Nations General Assembly (UNGA). (1948). Universal Declaration of Human Rights. UNGA res 217 A(III) (UDHR).

United Nations Educational, Scientific and Cultural Organization (UNESCO). (2021). Recommendation on the Ethics of Artificial Intelligence. SHS/BIO/PI/2021/1. https://unesdoc.unesco.org/ark:/48223/pf0000381137 Accessed 17 November 2022.

United Nations General Assembly (UNGA). (1992, August 12). Report of the United Nations Conference on Environment and Development (Rio de Janeiro, 3-14 June 1992). Annex I: Rio declaration on environment and development, A/CONF.151/26 (Vol. I).

United Nations General Assembly (UNGA). (1998, July 17). Rome Statute of the International Criminal Court (last amended 2010).

United Nations General Assembly (UNGA). (2005, December 16). Basic Principles and Guidelines on the Right to a Remedy and Reparation for Victims of Gross Violations of International Human Rights Law and Serious Violations of International Humanitarian Law. Adopted and proclaimed by General Assembly resolution 60/147, A/RES/60/147. https://www.ohchr.org/sites/default/files/2021-08/N0549642.pdf Accessed 15 November 2022.

United Nations General Assembly (UNGA). (2019, December 18). Seventy-fourth session. Agenda item 75. Responsibility of States for internationally wrongful acts. Resolution adopted by the General Assembly, A/RES/74/180. https://digitallibrary.un.org/record/3846830?ln=zh_CN Accessed 15 November 2022.

United Nations General Assembly (UNGA), Human Rights Council. (2011, March 21). Seventeenth session. Agenda item 3: Promotion and protection of all human rights, civil, political, economic, social and cultural rights, including the right to development. Report of the Special Representative of the Secretary General on the issue of human rights and transnational corporations and other business enterprises, John Ruggie. Guiding Principles on Business and Human Rights: Implementing the United Nations "Protect, Respect and Remedy" Framework. A/HRC/17/31. https://documents-dds-ny.un.org/doc/UNDOC/GEN/G11/121/90/PDF/G1112190.pdf?OpenElement Accessed 17 November 2022.

United Nations General Assembly (UNGA), Human Rights Council. (2013, October 8). Twenty-fourth session. Agenda item 3: Promotion and protection of all human rights, civil, political, economic, social and cultural rights, including the right to development. Resolution adopted by the Human Rights Council 24/16. The role of prevention in the promotion and protection of human rights. A/HRC/RES/24/16. Retrieved from https://ap.ohchr.org/documents/dpage_e.aspx?si=A/HRC/RES/24/16 Accessed 17 November 2022.

United Nations General Assembly (UNGA), Human Rights Council. (2015, July 16). Thirtieth session. Items 2 and 3 of the provisional agenda. Annual report of the United Nations High Commissioner for Human Rights. The Role of Prevention in the Promotion and Protection of Human Rights. Report of the Office of the United Nations High Commissioner for Human Rights. A/HRC/30/20. Retrieved from https://digitallibrary.un.org/record/801293?ln=en#record-files-collapse-header Accessed 17 November 2022.

United Nations General Assembly (UNGA), Human Rights Council. (2016, February 4). Thirty-first session. Agenda item 3: Promotion and protection of all human rights, civil, political, economic, social and cultural rights, including the right to development. Joint report of the Special Rapporteur on the rights to freedom of peaceful assembly and of association and the Special Rapporteur on extrajudicial, summary or arbitrary executions on the proper management of assemblies. A/HRC/31/66. Retrieved from https://undocs.org/Home/Mobile?FinalSymbol=A%2FHRC%2F31%2F66&Language=E&DeviceType=Desktop&LangRequested=False Accessed 17 November 2022.

United Nations General Assembly (UNGA), Human Rights Council. (2021, December 28). Forty-ninth session. 28 February–1 April 2022. Agenda item 3: Promotion and protection of all human rights, civil, political, economic, social and cultural rights, including the right to development. Rights of persons with disabilities, Report of the Special Rapporteur on the rights of persons with disabilities. A/HRC/49/52. https://www.undocs.org/Home/Mobile?FinalSymbol=A%2FHRC%2F49%2F52&Language=E&DeviceType=Desktop&LangRequested=False Accessed 17 November 2022.

United Nations International Covenant on Civil and Political Rights Human Rights Committee (UN ICCPRC). (2004, March 29). General Comment No 31: The Nature of the General Legal Obligation Imposed on States Parties to the Covenant. CCPR/C/21/Rev.1/Add. 13. https://digitallibrary.un.org/record/533996?ln=en#record-files-collapse-header Accessed 16 November 2022.

United Nations International Covenant on Civil and Political Rights Human Rights Committee (UN ICCPRC). (2019, September 3). General Comment No. 36: Article 6: Right to Life. CCPR/C/GC/36. https://digitallibrary.un.org/record/3884724?ln=en Accessed 16 November 2022.

United Nations International Law Commission (UN ILC). (1973, May 7-July 13). Report of the International Law Commission on the work of its twenty-fifth session. A/9010/Rev.1. Official Records of the General Assembly, Twenty-eighth session, Supplement No. 10.

United Nations International Law Commission (UN ILC). (2001a). Draft Articles on Prevention of Transboundary Harm from Hazardous Activities, with commentaries. Yearbook of the International Law Commission, 2001, vol. II, Part Two. https://legal.un.org/ilc/texts/instruments/english/commentaries/9_7_2001.pdf Accessed 16 November 2022.

United Nations International Law Commission (UN ILC). (2001b). Draft Articles on Responsibility of States for Internationally Wrongful Acts with commentaries. Yearbook of the International Law Commission, 2001, vol. II, Part Two. https://legal.un.org/ilc/texts/instruments/english/commentaries/9_6_2001.pdf Accessed 16 November 2022.

United Nations International Law Commission (UN ILC). (2001c). Draft Articles on Responsibility of States for Internationally Wrongful Acts, 2001. Supplement No. 10 (A/56/10), chp.IV.E.1.

United Nations International Law Commission (UN ILC). (2013). Report of the International Law Commission. A/68/10. Sixty-fifth session (6 May–7 June and 8 July–9 August 2013), Chapter XII: Other Decisions and Conclusions of the Commission. https://legal.un.org/ilc/reports/2013/english/chp12.pdf Accessed 16 November 2022.

United Nations International Law Commission (UN ILC). (2019, August 20). Draft Conclusions on Peremptory Norms of General International Law (Jus cogens). Report of the International Law Commission, Seventy-First Session, General Assembly Official Records, Supp No. 10 (A/74/10). https://legal.un.org/ilc/reports/2019/english/a_74_10_advance.pdf Accessed 16 November 2022.

United Nations (UN) Office at Geneva. (2020, July). All Drone Strikes in Self-Defence Should Go to Security Council, Argues UN Special Rapporteur. https://www.ungeneva.org/en/news-media/news/2020/07/all-drone-strikes-self-defence-should-go-security-council-argues. Accessed November 1, 2019.

United Nations Office for Disarmament (UNODA). (n.d.) Timeline of LAWS in the CCW. https://disarmament.unoda.org/timeline-of-laws-in-the-ccw/ Accessed 7 December 2023.

United Nations Security Council (UNSC). (2021, March 8). Letter dated 8 March 2021 from the Panel of Experts on Libya established pursuant to resolution 1973 (2011) addressed to the President of the Security Council. S/2021/229. Retrieved from https://undocs.org/S/2021/229 Accessed 17 November 2022.

United Nations (UN) Secretary-General. (2004). The rule of law and transitional justice in conflict and post-conflict societies. Report to the Security Council. UN Doc. S/2004/616.

United Nations (UN) Secretary-General. (2019, April 23). Responsibility of States for internationally wrongful acts: Compilation of decisions of international courts, tribunals and other bodies. Report of the Secretary-General. A/74/83. https://documents-dds-ny.un.org/doc/UNDOC/GEN/N19/117/91/PDF/N1911791.pdf?OpenElement Accessed 15 November 2022.

United States (U.S.) Delegation. (2018a, August 28). Human-Machine Interaction in the Development, Deployment and Use of Emerging Technologies in the Area of Lethal Autonomous Weapons Systems. Submitted by the United States. CCW/GGE.2/2018/

WP.4. https://reachingcriticalwill.org/images/documents/Disarmament-fora/ccw/2018/gge/documents/GGE.2-WP4.pdf Accessed 14 November 2022.

United States (U.S.) Delegation. (2018b, March 28). Humanitarian benefits of emerging technologies in the area of lethal autonomous weapon systems. Submitted by the United States of America. CCW/GGE.1/2018/WP.4. https://www.reachingcriticalwill.org/images/documents/Disarmament-fora/ccw/2018/gge/documents/GGE.1-WP4.pdf Accessed 14 November 2022.

United States (U.S.) Delegation. (2018c, April 13). US Statement as delivered by Ian R. McKay, CCW GGE. http://reachingcriticalwill.org/images/documents/Disarmament-fora/ccw/2018/gge/statements/13April_US.pdf Accessed 14 November 2022 Accessed 14 November 2022.

United States (U.S) Delegation. (2020, September 1). U.S. Commentaries on the Guiding Principles. https://documents.unoda.org/wp-content/uploads/2020/09/20200901-United-States.pdf Accessed 14 November 2022.

United States, Department of Defense (US DoD). (2012, November 21). Directive n. 3000.09. https://www.esd.whs.mil/portals/54/documents/dd/issuances/dodd/300009p.pdf Accessed 14 November 2022.

United States, Department of Defense (US DoD). (2015). Law of War Manual. https://dod.defense.gov/Portals/1/Documents/pubs/DoD%20Law%20of%20War%20Manual%20-%20June%202015%20Updated%20Dec%202016.pdf?ver=2016-12-13-172036-190 Law of War Manual.

United States, Department of Defense (US DoD). (2023, January 25). Directive n. 3000.09. https://media.defense.gov/2023/Jan/25/2003149928/-1/-1/0/DOD-DIRECTIVE-3000.09-AUTONOMY-IN-WEAPON-SYSTEMS.PDF Accessed 10 March 2023.

United States (US). (n.d.). U.S. Supreme Court advanced search for opinions and court orders. Retrieved from https://www.supremecourt.gov/search_center.aspx.

Verri, P. and Markee, E. (1992). Dictionary of the International Law of Armed Conflict. International Committee of the Red Cross. https://www.icrc.org/en/publication/0453-dictionary-international-law-armed-conflict. Accessed November 13, 2022.

Vilmer, J.-B.J. (2021). A French Opinion on the Ethics of Autonomous Weapons. War on the Rocks. https://warontherocks.com/2021/06/the-french-defense-ethics-committees-opinion-on-autonomous-weapons/ Accessed 17 November 2022.

Vöneky, S. (2008). Analogy in International Law. Max Planck Encyclopedias of Public International Law [MPEPIL]. Published under the auspices of the Max Planck Institute for Comparative Public Law and International Law under the direction of Professor Anne Peters (2021–) and Professor Rüdiger Wolfrum (2004–2020). https://opil.ouplaw.com/view/10.1093/law:epil/9780199231690/law-9780199231690-e1375#:~:text=1%20Generally%20speaking%2C%20the%20use,of%20legal%20reasoning%20by%20analogy Accessed 17 November 2022.

Wallach, W. (2015). A Dangerous Master: How to Keep Technology from Slipping beyond Our Control. Basic Books, a member of the Perseus Books Group.

Wareham, M. (2019, March 27). Statement on options for future work by the campaign to stop killer robots, CCW meeting on lethal autonomous weapons systems. Human Rights Watch for the Campaign to Stop Killer Robots. https://www.hrw.org/node/330486/printable/print Accessed 14 November 2022.

Wareham, M. (2020). Stopping killer robots: Country positions on banning fully autonomous weapons and retaining human control. Human Rights Watch 2020. https://www.hrw.org/sites/default/files/media_2021/04/arms0820_web_1.pdf Accessed 12 November 2022.

Weber, J. and Suchman, L. (2016). Human–machine autonomies. *In:* N. Bhuta, S. Beck, R. Geiβ, H. Liu & C. Kreβ (Eds.), Autonomous Weapons Systems: Law, Ethics, Policy (pp. 75-102). Cambridge: Cambridge University Press. doi:10.1017/CBO9781316597873.004 .

Yampolskiy, R.V. (2020). Unpredictability of AI: On the impossibility of accurately predicting all actions of a smarter agent. Journal of Artificial Intelligence and Consciousness, 7(1), 109-118. https://doi.org/10.1142/S2705078520500034.

Yang, Q. (2018). The Fourth Revolution. UNESCO Courier, (3). https://unesdoc.unesco.org/ark:/48223/pf0000265246 Accessed 19 May 2019.

Legally Binding Instruments

Additional Protocol on Prohibitions or Restrictions on the Use of Incendiary Weapons (Protocol III, on Incendiary Weapons). (1983). 1342 UNTS 137. United Nations (UN).

Additional Protocol to the Convention on Prohibitions or Restrictions on the Use of Certain Conventional Weapons which may be deemed to be Excessively Injurious or to have Indiscriminate Effects (CCW) (Protocol IV, on Blinding Laser Weapons). (1998). 1380 UNTS 370. United Nations. (U.N).

Convention Against Torture and Other Cruel, Inhuman or Degrading Treatment or Punishment. (1987). 1465 UNTS 85. United Nations.

Convention for the Protection of Individuals with Regard to the Automatic Processing of Individual Data. (1981). ETS 108 No. 108. https://rm.coe.int/1680078b37 Accessed 27 August 2023.

Convention on Civil Liability for Damage Resulting from Activities Dangerous to the Environment. (1993). ETS No. 150. Council of Europe. https://rm.coe.int/168007c079 Accessed 27 August 2023.

Convention on International Liability for Damage Caused by Space Objects. (1972). 961 UNTS 187. United Nations.

Convention on Civil Liability for Oil Pollution Damage (CLC). (1969). UNTS 1975. International Maritime Organization.

Convention on the Law of the Sea (UNCLOS). (1982). 1833 UNTS. United Nations.

Convention on the Prohibition of the Development, Production, Stockpiling and Use of Chemical Weapons and on their Destruction (Chemical Weapons Convention). (1997). 1975 UNTS 45. United Nations.

Convention on the Prohibition of the Use, Stockpiling, Production and Transfer of Anti-Personnel Mines and on their Destruction (Anti-Personnel Landmines Convention) (1997). 2056 UNTS 211. United Nations (UN).

Convention on Prohibitions or Restrictions on the Use of Certain Conventional Weapons Which May be Deemed to be Excessively Injurious or to Have Indiscriminate Effects (and Protocols) (As Amended on 21 December 2001) (CCW). (1983). 1342 UNTS 137. United Nations.

Geneva Convention for the Amelioration of the Condition of the Wounded and Sick in Armed Forces in the Field (First Geneva Convention). (1949). 75 UNTS 31. International Committee of the Red Cross (ICRC).

Geneva Convention for the Amelioration of the Condition of the Wounded and Sick and Shipwrecked Members of Armed Forces at Sea (Second Geneva Convention). (1949). 75 UNTS 61. International Committee of the Red Cross (ICRC).

Geneva Convention Relative to the Treatment of Prisoners of War (Third Geneva

Convention). (1949). 75 UNTS 135. International Committee of the Red Cross (ICRC).

Geneva Convention Relative to the Protection of Civilian Persons in Time of War (Fourth Geneva Convention). (1949). 75 UNTS 287. International Committee of the Red Cross (ICRC).

Hague Convention IV Respecting the Laws and Customs of War on Land and Its Annex: Regulations Concerning the Laws and Customs of War on Land. (Hague Convention IV) (1910). International Conferences (The Hague).

International Covenant on Civil and Political Rights (ICCPR). (1976). 999 UNTS 171. United Nations.

Protocol Additional to the Geneva Conventions of 12 August 1949, and relating to the Protection of Victims of International Armed Conflicts (Additional Protocol I). (1977). 1125 UNTS 3. International Committee of the Red Cross (ICRC).

Protocol amending the Convention for the Protection of Individuals with regard to Automatic Processing of Personal Data. (2018). ETS No. 223. Council of Europe. https://rm.coe.int/16808ac918 Accessed 27 August 2023.

Protocol on Prohibitions or Restrictions on the Use of Mines, Booby-Traps and Other Devices as amended on 3 May 1996 (Protocol II, as amended on 3 May 1996) (Protocol II, on the Use of Mines, Booby-Traps and Other Devices). (1998). Annexed to the Convention on Prohibitions or Restrictions on the Use of Certain Conventional Weapons which may be deemed to be Excessively Injurious or to have Indiscriminate Effects. 2048 UNTS 93. United Nations (UN).

Second Protocol to the Hague Convention of 1954 for the Protection of Cultural Property in the Event of Armed Conflict. (1999). UNESCO.

Treaty on Principles Governing the Activities of States in the Exploration and Use of Outer Space, including the Moon and Other Celestial Bodies. (1967). 8843 UNTS. United Nations.

Treaty on the Prohibition of Nuclear Weapons. (2017). Adopted 7 July 2017, entered into force 22 January 2021. Document: A/CONF.229/2017/8 and C.N.478.2020. TREATIES-XXVI.9. United Nations.

Treaty on the Non-Proliferation of Nuclear Weapons (NPT). (1970). 729 UNTS 161. United Nations.

Vienna Convention on the Law of Treaties. (1969). Adopted 23 May 1969, entered into force 27 January 1980. Published in UNTS vol. 1155, p. 331. United Nations.

Cases

Inter-American Court of Human Rights (IACHR). (1989). Velásquez Rodríguez Case, Judgment.

International Criminal Court (ICC). (2018). Prosecutor v. Germain Katanga, Decision on the Matter of the Transgenerational Harm Alleged by Some Applicants for Reparations Remanded by the Appeals Chamber in its Judgment of 8 March 2018, 19 July 2018. ICC-01/04-01/07. Trial Chamber II.

International Court of Justice (ICJ). (1949). Corfu Channel Case, Judgment. ICJ Rep 1949.

International Court of Justice (ICJ). (1969). North Sea Continental Shelf Case (Federal Republic of Germany v Netherlands) Merits. ICJ Rep 1969.

International Court of Justice (ICJ). (1970). Case Concerning the Barcelona Traction, Light and Power Company, Limited (Belgica v. Espanha) New Application: 1962 - Judgment. ICJ Rep 1970.

International Court of Justice (ICJ). (1986). Military and Paramilitary Activities in and against Nicaragua (Nicaragua v. United States of America) Judgment. ICJ Rep 1986.

International Court of Justice (ICJ). (1996). Legality of the Threat or Use of Nuclear Weapons, Advisory Opinion. ICJ Rep 1996.

International Court of Justice (ICJ). (2001). LaGrand Case (Germany v United States), Judgment, Jurisdiction, Admissibility, Merits. ICJ Rep 466.

International Court of Justice (ICJ). (2002). Land and Maritime Boundary between Cameroon and Nigeria Case (Cameroon v. Nigeria: Equatorial Guinea intervening) Judgment. ICJ Rep 303.

International Court of Justice (ICJ). (2004). Legal Consequences of the Construction of a Wall in the Occupied Palestinian Territory, Advisory Opinion. ICJ Rep 136.

International Court of Justice (ICJ). (2007). Application of the Convention on the Prevention and Punishment of the Crime of Genocide (Bosnia and Herzegovina v Serbia and Montenegro) (Bosnian Genocide case). Judgment. ICJ Rep 43.

International Court of Justice (ICJ). (2008). Certain Questions of Mutual Assistance in Criminal Matters (Djibouti v France) Judgment. ICJ Rep 177.

International Court of Justice (ICJ). (2010). Pulp Mills on the River Uruguay (Argentina v. Uruguay), Judgment. ICJ Rep 2010.

International Court of Justice (ICJ). (2012). Jurisdictional Immunities of the State (Germany v Italy) Judgment. ICJ Rep. 434.

International Court of Justice (ICJ). (2015). Certain Activities Carried Out by Nicaragua in the Border Area (Costa Rica v. Nicaragua) and Construction of a Road in Costa Rica along the San Juan River (Nicaragua v. Costa Rica) Judgment. ICJ Rep 2015.

International Military Tribunal Sitting at Nuremberg. (1946). Trial of German Major War Criminals. Proceedings. Part 22 (1946). https://avalon.law.yale.edu/imt/09-30-46.asp Accessed 16 November 2022.

International Tribunal for the Law of the Sea (ITLOS). (2011). Responsibilities and obligations of States with respect to activities in the Area, Advisory Opinion. Seabed Disputes Chamber. List of cases: No. 1750 ILM 458 https://www.itlos.org/fileadmin/itlos/documents/cases/case_no_17/17_adv_op_010211_en.pdf Accessed 17 November 2022.

Permanent Court of International Justice (PCIJ). (1927). S.S. 'Lotus' (France v Turkey), Judgment, Judgment No 9. Series A No 10.

Report of International Arbitral Awards. (1990). Case concerning the difference between New Zealand and France concerning the interpretation or application of two agreements concluded on 9 July 1986 between the two States and which related to the problems arising from the Rainbow Warrior Affair Decision of 30 April 1990. Vol. XX, pp. 215-284.

United Kingdom (UK) Court of Appeal. (2021). Thaler v Comptroller General of Patents Trade Marks and Designs. EWCA Civ 1374 https://www.bailii.org/ew/cases/EWCA/Civ/2021/1374.pdf Accessed 16 November 2022.

Index

M

N

O

P

R

S

T

For Product Safety Concerns and Information please contact our EU representative GPSR@taylorandfrancis.com
Taylor & Francis Verlag GmbH, Kaufingerstraße 24, 80331 München, Germany

www.ingramcontent.com/pod-product-compliance
Lightning Source LLC
LaVergne TN
LVHW010553110826
845149LV00003B/647

* 9 7 8 1 0 3 2 6 9 2 3 4 0 *